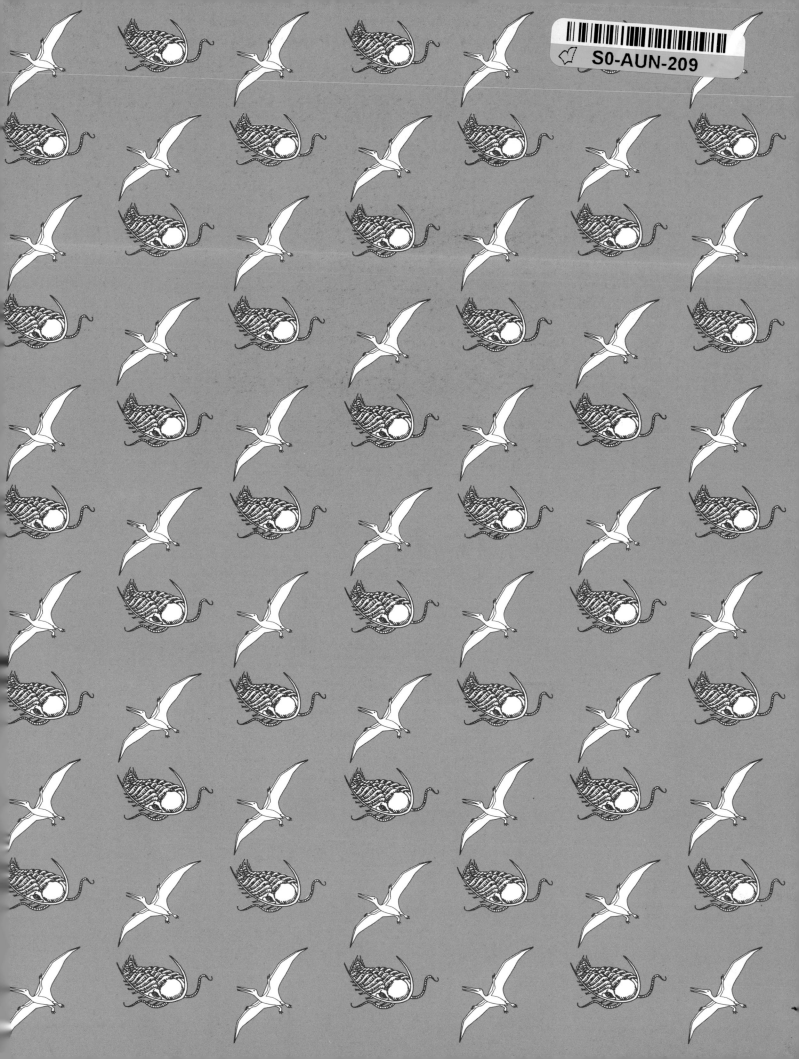

EYEWITNESS BOOKS

PREHISTORIC
LIFE

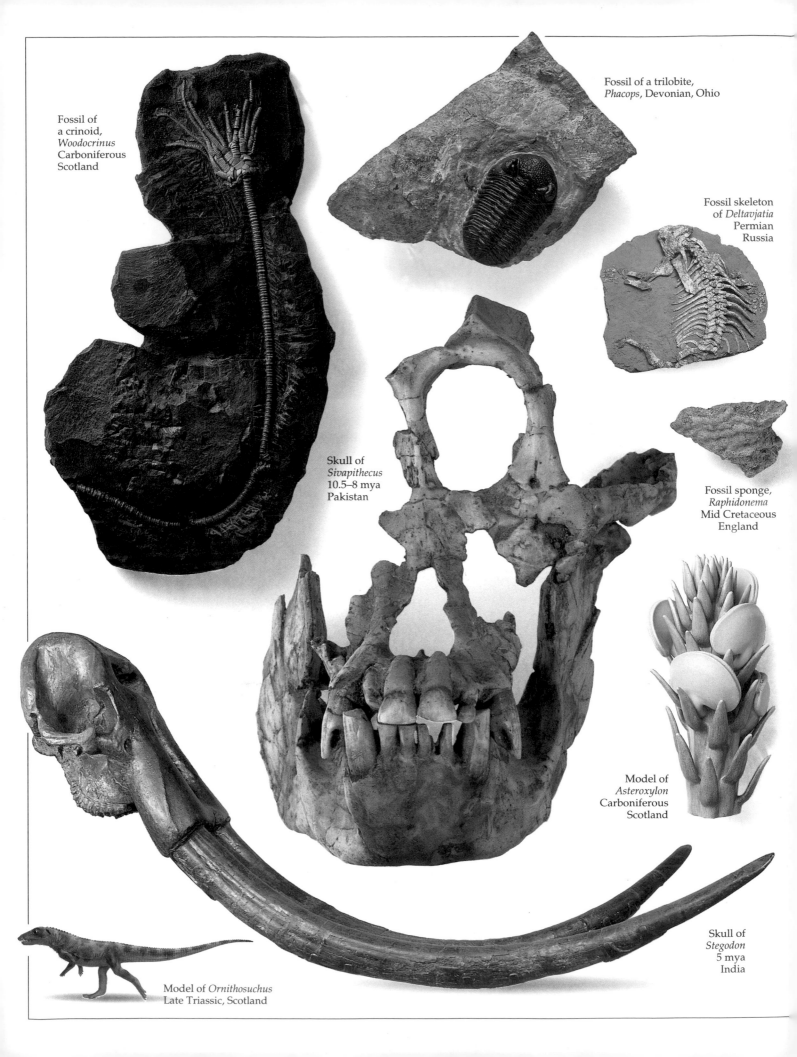

Fossil of a trilobite, *Phacops*, Devonian, Ohio

Fossil of a crinoid, *Woodocrinus* Carboniferous Scotland

Fossil skeleton of *Deltavjatia* Permian Russia

Skull of *Sivapithecus* 10.5–8 mya Pakistan

Fossil sponge, *Raphidonema* Mid Cretaceous England

Model of *Asteroxylon* Carboniferous Scotland

Model of *Ornithosuchus* Late Triassic, Scotland

Skull of *Stegodon* 5 mya India

Araucaria cone
Jurassic, Argentina

Model of
Homo habilis

EYEWITNESS BOOKS

PREHISTORIC LIFE

Written by
WILLIAM LINDSAY

Photographed by
HARRY TAYLOR

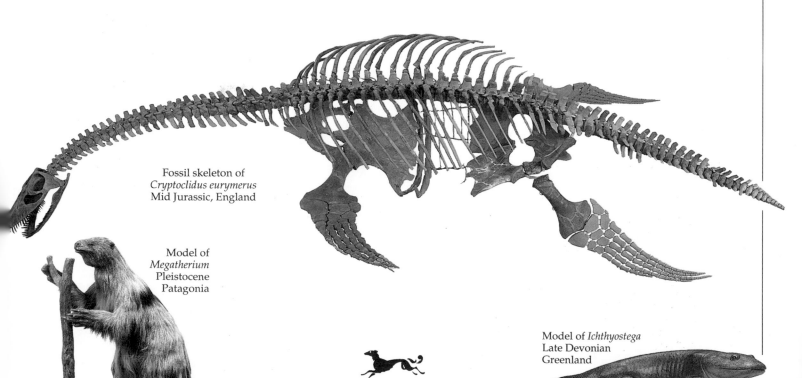

Fossil skeleton of
Cryptoclidus eurymerus
Mid Jurassic, England

Model of
Megatherium
Pleistocene
Patagonia

Model of *Ichthyostega*
Late Devonian
Greenland

ALFRED A. KNOPF • NEW YORK

Model of
Arthropleura
Carboniferous
Scotland

Top view
of skull of
Simolestes
Mid Jurassic
England

Model of
Cothurnocystis
Late Ordovician
Scotland

Ediacara
Precambrian
Australia

A DORLING KINDERSLEY BOOK

Project editor Marion Dent
Art editor Vicky Wharton
Managing editor Simon Adams
Managing art editor Julia Harris
Researcher Céline Carez
Picture research Deborah Pownall
Production Catherine Semark
Editorial consultant Dr Robin Cocks, Head of
Palaeontology, Natural History Museum (London)

Special thanks to National Museums of Scotland
(Edinburgh), Hunterian Museum (University of Glasgow),
Natural History Museum (London), Sedgwick and
Zoology Museums (University of Cambridge)

This is a Borzoi Book published by
Alfred A. Knopf, Inc.

This Eyewitness ™ Book has been conceived by Dorling
Kindersley Limited and Editions Gallimard

First American edition, 1994
Copyright © 1994 by Dorling Kindersley Limited, London.
Text copyright © 1994 by William Lindsay
All rights reserved under International and Pan-American
Copyright Conventions. Published in the United States by
Alfred A. Knopf, Inc., New York. Distributed by Random
House, Inc., New York. Published in Great Britain by
Dorling Kindersley Limited, London.

Manufactured in Singapore
0 9 8 7 6 5 4 3 2

Library of Congress Cataloging in Publication Data
Lindsay, William.
Prehistoric life / written by William Lindsay — 1st
American ed.
p. cm. — (Eyewitness books)
Includes index.
1. Fossils—Juvenile literature. 2. Evolution (Biology)
—Juvenile literature. [1. Fossils. 2. Evolution] I. Title
QE714.5.L53 1994
560—dc20   93-32076
ISBN 0-679-86001-0
ISBN 0-679-96001-5 (lib. bdg.)

Color reproduction by Colourscan, Singapore
Printed in Singapore by Toppan

Fossil of *Archaeopteris*
Late Devonian, Ireland

Fossil skeleton of *Naso
rectifrons*, Late Eocene, Italy

Fossil skeleton
of *Procolophon*
Early Triassic
South Africa

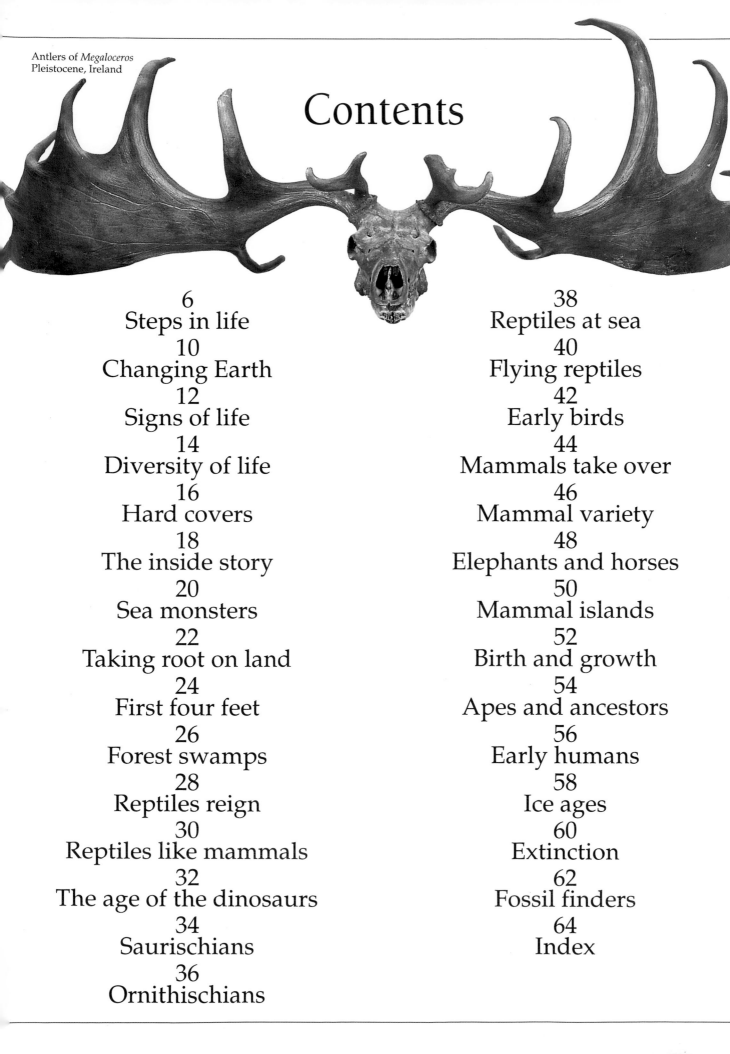

Antlers of *Megaloceros*
Pleistocene, Ireland

Contents

Steps in life

How DID LIFE BEGIN? No one knows for sure. Earth itself formed 4.5 billion years ago when clouds of particles drew together under great heat and pressure. Then, scientists believe, chemicals in the atmosphere combined about 3.5 billion years ago and set off an explosion of life. Fossils – the hardened remains of dead plants and animals – show that different life-forms evolved at different stages of Earth's geological history. At the bottom of the next few pages is a timeline of that history. It is divided into great eras lasting millions of years. Each era is divided into distinct periods, or even shorter timespans called epochs.

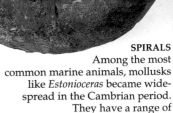
Cothurnocystis
Ordovician

SEA-FLOOR DWELLER
Many animals build a supporting scaffold and protective shell for their bodies. Their shells were hard and are easily preserved as fossils. Sea-floor dwelling *Cothurnocystis* (pp. 16–17) found in the U.K. may be related to all backboned animals.

Metaldetes
Cambrian

CONE-SHAPED
Metaldetes was a sponge-like marine animal (pp. 16–17). It had a double-walled skeleton to support its body.

CURIOUS FOSSIL
Mawsonites from Australia (pp. 12–13) is about 700 million years old. It may be one of the earliest multicelled animals, such as a jellyfish.

Mawsonites,
Precambrian, Australia

ROCKY MAT
Among the oldest, most primitive fossils, 3.4-billion-year-old stromatolites (pp. 12–13) were some of the first signs of life on Earth.

Estonioceras
Early Ordovician
Estonia

Collenia
(a stromatolite)
Precambrian, U.S.

WEIRD ANIMAL
Whole communities of animals were fossilized together, as in Canada's Burgess Shale (pp. 14–15). Feather-shaped *Thaumaptilon* is from Burgess.

SPIRALS
Among the most common marine animals, mollusks like *Estonioceras* became widespread in the Cambrian period. They have a range of structures and shapes.

BUBBLING BEDS
Early Earth's surface was a hotbed of bubbling volcanoes and fiery lava flows (pp. 10–11). As the Earth cooled, lavas hardened as rock.

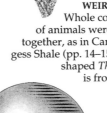

PRECAMBRIAN PERIOD
Before 570 mya (mya = millions of years ago)

CAMBRIAN PERIOD
570–510 mya

ORDOVICIAN PERIOD
510–439 mya

A SLOW START
The first era of geological history was the Precambrian, lasting some 4 million years. Although very few rocks are left from the first million years of the Precambrian, evidence suggests that early Earth had no oxygen in its air and no life on its land. The continents of today were probably joined as one or a few landmasses. The first simple life-forms – single-celled bacteria and algae – evolved in the shallow primeval oceans of the Precambrian about 3.5 billion years ago. Another million years passed before an oxygen-rich atmosphere developed on Earth, and more complex, multicelled and soft-bodied organisms evolved.

HARD SHELLS
Cambrian animals developed body frames and shelly covers for protection. Trilobites (creatures with segmented armor) and brachiopods (tiny, shelled animals) were common. There was one large land mass, Gondwanaland. North America, Europe, and Greenland existed as smaller, separate continents.

EARLY FISH
The first fishlike creatures developed. They had no backbones at first, and no jaws. Worms, snails, and jellyfish became common. The Ordovician was a time of shrinking oceans and ice ages – periods when temperatures dropped and glaciers covered the land. Gondwanaland and the other continents were moving slowly together.

PRECAMBRIAN ERA   4,600–570 mya   PALEOZOIC ERA   570–245 mya

Calymene blumenbachii – a trilobite curled up for protection from predators Silurian, U.K.

EARLY PLANTS
Plants became dominant during the late Devonian period. In the Carboniferous period, giant club-mosses, such as *Lepidodendron* (pp. 24–25), formed huge swamp forests.

CURLED-UP CALYMENE
Although hard-shelled, trilobites (pp. 16–17) were a highly mobile group of sea-bottom animals. Their external body skeletons were jointed and segmented. Some, like *Calymene*, could roll up for protection from predators.

Cyathocrinites Silurian U.K.

Model of *Westlothiana* Carboniferous Scotland

Derbyia Permian India

CRINOIDS
Plantlike animals called crinoids, such as *Cyatho-crinites*, were so common in places that layers of limestone are composed completely of their fossilized remains.

Engraving of *Cephalaspis* Silurian

FIRST REPTILE
Four-legged animals took a major step forward when reptiles appeared. Unlike their amphibian ancestors, reptiles– such as the earliest known *Westlothiana* (pp. 28–29) – did not need an aquatic environment for survival.

PROMINENT SHELLFISH
Once brachiopods (pp. 18–19) – tiny, shelled creatures, such as *Derbyia* – were the most prominent shellfish. Their place was taken by today's common mollusks.

JAWLESS FISH
An internal bony skeleton was revolutionary in animal development. Jawless fish (pp. 18–19) like *Cephalaspis* appeared by the end of the Silurian period.

SILURIAN PERIOD
439–409 mya

DEVONIAN PERIOD
409–363 mya

CARBONIFEROUS PERIOD
363–290 mya

PERMIAN PERIOD
290–245 mya

INVASION OF LAND
Until the Silurian period, life on Earth had been confined to the oceans, with no organisms surviving out of water. During the Silurian, mountain ranges were forming across Scandinavia, Scotland, and the coast of North America. Plants made their first appearance on land and giant scorpions hunted in the seas.

SWARMING SEAS
The invasion of land by animals continued in the Devonian period. While sharks and jawed fish became active predators in the seas, amphibians made their move onto land. As oceans narrowed, the giant continent of Gondwanaland closed in on Europe, North America, and Greenland.

THE FIRST REPTILES
During the Carboniferous period, the Earth's continents formed a single land-mass, changing the environment for many forms of life. In the Northern Hemisphere, tropical climates produced vast, tree-filled swamps, preserved today as coal. Reptiles first appeared, able to lay eggs out of water.

MASS EXTINCTION
The continents fused together and moved as one land-mass (Pangaea). Ice sheets covered South America, Antarctica, Australia, and Africa, locking up water and lowering the sea level. Dry, desert conditions existed in the north. The Permian period ended with mass extinctions on the largest scale ever.

Continued on next page

Model of
Megazostrodon
Late Triassic
South Africa

Life continues

After the mass extinctions at the end of the Permian, 245 million years ago (mya), the Mesozoic era began, marked by huge geological changes and a wealth of new plant and animal life. Mammal-like reptiles became the most prominent animals. When they died out, dinosaurs appeared and dominated Earth. Mammals evolved and survived the extinction of the dinosaurs, as did the dinosaurs' descendants, the birds. Evolution took many twists and turns. Some paths led to the ancestors of today's animals – others led to dead ends: species only on Earth a short time before dying out.

FIRST FLYING MAMMALS
Palaeochiropteryx, a bat from Messel, Germany, looks very similar to today's bats.

SMALL MAMMALS
Megazostrodon developed from mammal-like reptiles near the end of the Triassic period. Unlike the cold-blooded reptiles, mammals maintained body heat with fur and with rapid processing of food for energy (pp. 44–51). Born live, mammal young fed on mother's milk.

Engraving of
Archaeopteryx
Jurassic
Germany

FIRST BIRD
From among a group of small, two-legged, meat-eating dinosaurs came the first birds. About 150 mya, *Archaeopteryx* (pp. 42–43), still with reptile teeth, tail, and fingers, flew after its insect prey.

LONG TAIL
Leptictidium, with an 8-in (20-cm) long tail, was an Eocene omnivore. Remains of plants, insects, and lizards have been found fossilized in the stomach area of this fast hunter. The long tail was a counterbalance for its body during the chase.

LARGEST ANIMALS
Dinosaurs (pp. 32–37) lived on Earth for a long time, from the Triassic period to the end of the Cretaceous period, when they became extinct (pp. 60–61).

Model of fearsome
Tyrannosaurus rex
Late Cretaceous
North America
and China

Reconstruction
of *Leptictidium*
Eocene
Germany

Stenopterygius
Jurassic
Germany

TINY HORSE
Hyracotherium, the first known horse, appeared. It was only the size of a dog and ran on four-toed feet (pp. 48–49).

MARINE REPTILES
Not all reptiles were land-based. Ichthyosaurs (pp. 38–39) made their home at sea. They did not lay eggs on land but gave birth (pp. 52–53) to live young at sea.

Hyracotherium
Late Paleocene to
Early Eocene
North America
and Europe

JURASSIC PERIOD
208–146
mya

CRETACEOUS PERIOD
146–65
mya

TRIASSIC PERIOD
245–208 mya

TERTIARY PERIOD
65–1.64 mya

AGE OF REPTILES
Pangaea drifted northward. Deserts formed. Southern ice sheets melted, and cracks appeared in Pangaea. Many reptiles evolved on land, including dinosaurs, while others took to the air or lived in the sea.

REIGN OF THE DINOSAURS
Dinosaurs spread across the lands, ichthyosaurs hunted in the seas, and pterosaurs dominated in the air. Pangaea broke up as North America drifted away from Africa, South America from Antarctica and Australia.

DINOSAURS DIE OUT
Milder climates appeared. India drifted away from Africa toward Asia. Dinosaurs died out at the end of this period, along with many other species. The extinctions were possibly due to a meteorite hitting Earth.

PALEOCENE EPOCH
65–56.5 mya
Mammals rapidly took over after the demise of the dinosaurs. Continents were now separate. Africa moved north on a collision course with Europe, while North America drifted eastward.

EOCENE EPOCH
56.5–35.5 mya
Swimming and flying mammals established many new habitats for themselves. India joined with Asia. The Atlantic Ocean separated Europe and North America. South America was on its own.

MESOZOIC ERA   245 mya – 65 mya

CENOZOIC ERA   65 mya – present

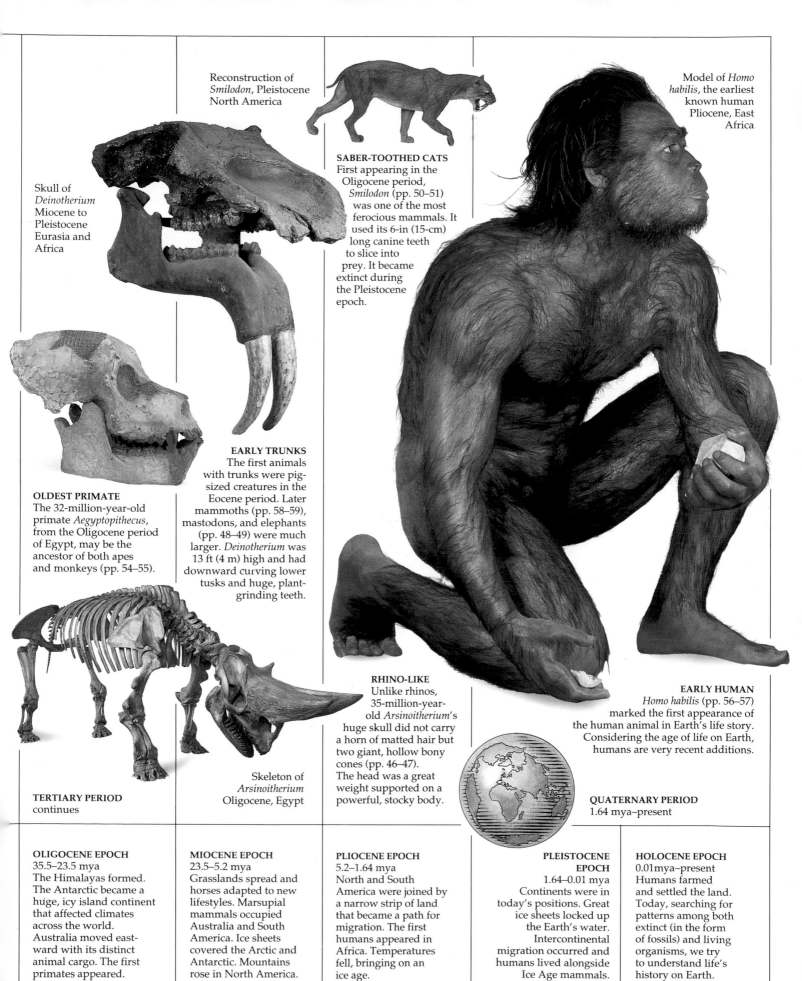

Reconstruction of *Smilodon*, Pleistocene North America

Model of *Homo habilis*, the earliest known human Pliocene, East Africa

Skull of *Deinotherium* Miocene to Pleistocene Eurasia and Africa

SABER-TOOTHED CATS
First appearing in the Oligocene period, *Smilodon* (pp. 50–51) was one of the most ferocious mammals. It used its 6-in (15-cm) long canine teeth to slice into prey. It became extinct during the Pleistocene epoch.

OLDEST PRIMATE
The 32-million-year-old primate *Aegyptopithecus*, from the Oligocene period of Egypt, may be the ancestor of both apes and monkeys (pp. 54–55).

EARLY TRUNKS
The first animals with trunks were pig-sized creatures in the Eocene period. Later mammoths (pp. 58–59), mastodons, and elephants (pp. 48–49) were much larger. *Deinotherium* was 13 ft (4 m) high and had downward curving lower tusks and huge, plant-grinding teeth.

RHINO-LIKE
Unlike rhinos, 35-million-year-old *Arsinoitherium*'s huge skull did not carry a horn of matted hair but two giant, hollow bony cones (pp. 46–47). The head was a great weight supported on a powerful, stocky body.

Skeleton of *Arsinoitherium* Oligocene, Egypt

EARLY HUMAN
Homo habilis (pp. 56–57) marked the first appearance of the human animal in Earth's life story. Considering the age of life on Earth, humans are very recent additions.

TERTIARY PERIOD continues

QUATERNARY PERIOD 1.64 mya–present

OLIGOCENE EPOCH
35.5–23.5 mya
The Himalayas formed. The Antarctic became a huge, icy island continent that affected climates across the world. Australia moved eastward with its distinct animal cargo. The first primates appeared.

MIOCENE EPOCH
23.5–5.2 mya
Grasslands spread and horses adapted to new lifestyles. Marsupial mammals occupied Australia and South America. Ice sheets covered the Arctic and Antarctic. Mountains rose in North America.

PLIOCENE EPOCH
5.2–1.64 mya
North and South America were joined by a narrow strip of land that became a path for migration. The first humans appeared in Africa. Temperatures fell, bringing on an ice age.

PLEISTOCENE EPOCH
1.64–0.01 mya
Continents were in today's positions. Great ice sheets locked up the Earth's water. Intercontinental migration occurred and humans lived alongside Ice Age mammals.

HOLOCENE EPOCH
0.01mya–present
Humans farmed and settled the land. Today, searching for patterns among both extinct (in the form of fossils) and living organisms, we try to understand life's history on Earth.

Changing Earth

FOR ALMOST A BILLION YEARS, nothing lived on Earth. Nothing wriggled or ran, flew or swam. That ancient, lifeless Earth was very different from the world today. But as the early planet was constantly evolving – as rocks formed, oceans spread, continents shifted, mountains rose, earthquakes and volcanoes shook and rattled the surface, and climates changed – the chance was created for life. The evidence of that life is today preserved as fossils found on the Earth's rocky surface.

HOT ROCKS
Volcanoes occur at weak, thin points in the Earth's crust. Molten lava is poured out of cracks and solidifies as it cools. Ash and hot gases are thrown into the air and the ash falls to form a volcanic cone. In the early stages of the Earth's formation, the world was a hot, molten mass.

Sedimentary sandstone is made of eroded quartz grains

Metamorphic marble was sedimentary limestone

Igneous granite solidified deep underground

Basalt is a common volcanic igneous rock

BUILDING BLOCKS
Igneous rocks form from molten rock material deep in the Earth and at the surface. Rock particles eroded by wind and water form sedimentary layers in rivers, seas, and lakes. Temperature and pressure can transform both igneous and sedimentary rocks into new, metamorphic rocks.

WATERY GRAVES
Rivers build thick sediment layers of sand and mud on flood plains and deltas. Reaching the sea, the sediment sinks to the sea floor. Quickly buried, animal and plant remains may be preserved as sediments turn to rock.

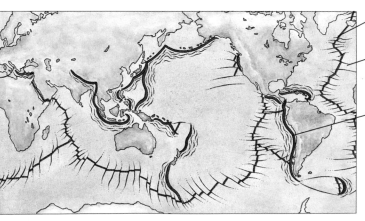

Volcanic ridge

Fault line

Plate-edge trench

The "part" of a trilobite

MOVING PLATES
Earth's surface is made of large interlocking plates. New molten rock rises from volcanic ridges along the ocean floor and adds to the plates. As the plates grow and spread, their edges collide, and sink into trenches, causing earthquakes and volcanoes. When continents carried on top of the plates collide, huge mountain ranges like the Himalayas in Asia may be created.

FOSSIL EVIDENCE
Fossils are evidence for ancient animal and plant life. They may preserve an organism's detailed inner structure as well as its outer shape. Flowers, feathers, and even footprints can be fossilized. Converting buried organisms into these stony replicas takes millions of years as organic material is destroyed and minerals in the rocks slowly fill in microscopic spaces. Sometimes the buried fossil is destroyed completely in the rock, leaving a natural mold. The space left behind is filled with more rock material, producing a natural and accurate cast of the fossil's shape.

How a fossil is formed

Decaying (1) *Procolophon* was covered in silt, sediment swept in by shallow streams (2). Burial must have been rapid since the skeleton was not broken up, although the flesh rotted. Over millions of years the skeleton was buried deep underground (3). Under pressure sands became stone, in which chemicals turned *Procolophon*'s bones into fossil. Erosion brought the fossil back to the surface (4).

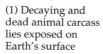

(1) Decaying and dead animal carcass lies exposed on Earth's surface

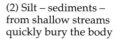

(2) Silt – sediments – from shallow streams quickly bury the body

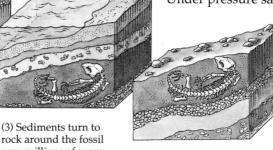

(3) Sediments turn to rock around the fossil over millions of years

(4) Fossilized skeleton is exposed at the surface

"Counterpart" of a trilobite, *Flexicalymene caractaci* Ordovician, England

Procolophon trigoniceps Early Triassic South Africa

SMALL REPTILE

Procolophon was a small reptile that lived during the Early Triassic period. Its fossils are well known from South Africa's Karroo Basin. Complete skeletons and skulls, perfectly preserved as white fossil bone, are buried in a red silty rock. This red color comes from iron minerals that hold the quartz grains together. It shows that the sediment was exposed to air and not continuously buried underwater, since iron rusts only in air.

HOT AND COLD

Rocks preserve clues of climatic conditions that help to explain the shifting continents. Fossils of the large-leaved *Glossopteris* are found in abundance in Antarctic Permian rocks. During this period, the Antarctic, part of a much larger continental mass, was not at the South Pole. Its climate was much warmer than today.

Fossil leaf

Glossopteris, Permian NSW, Australia

PART AND COUNTERPART

Splitting apart the rock reveals the positive "part" of the fossil and the negative "counterpart" – the natural mold. This trilobite's overall shape has been preserved, but the hard fossil skeleton was destroyed over many years. Fossils are often flattened by the pressure of overlying rock and may be cracked and broken into many pieces.

OLD AGE

Evidence from meteorites and moon rocks show that the Earth is 4.5 billion years old. The oldest rocks on Earth, about 3.75 billion years old, are found in Greenland and are metamorphic. Some of these rocks were originally sediments, laid down at an even earlier stage.

Signs of life

THE OLDEST SIGNS OF LIFE discovered so far on Earth are microscopically small strands of single cells from 3.4–3.3 billion-year-old rocks in Australia and South Africa. These are the remains of cyanobacteria (blue-green algae) that appeared a billion years after the Earth formed. Cyanobacteria are prokaryotes – simple, single-celled microorganisms. Many of them can live only in a low oxygen environment. Plants, animals, and fungi are all eukaryotes; their cells are divided into separate compartments with each part having a special job to do. Eukaryotes first appeared about two billion years ago. They may have begun as a group of prokaryotes that joined together. Before this time there was not enough oxygen in the atmosphere for eukaryotes to survive and no ozone layer to protect them from the sun's harmful ultraviolet light.

ALGAE MOUNDS
Mounds of stromatolites grow in the shallow, clear water of Shark Bay in Western Australia. Stromatolites are layers of intertwined blue-green algae that grow by photosynthesis, absorbing carbon dioxide and giving off oxygen.

Each layer shows a period of growth

GROWTH LINES
Mats of blue-green algae are sticky and trap sediment particles. Cemented together over the years, layers build up to make the lumpy mounds of rock that are found as fossilized stromatolites. Three-billion-year-old stromatolites have been found in South Africa.

Scientists think Spriggina's *"head" was actually an attachment for a frond*

Skakoper cryptozoan, a stromatolite Proterozoic Minnesota

Segment was a body part (leaf of frond)

Dickinsonia *from Australia's Ediacara grew by adding new segments to its body*

An Ediacara *fossil, probably buried by sand where it lived during the Precambrian in South Australia*

Two jellyfish from Ediacara seem to have been fossilized side by side

SOFT IMPRESSIONS
In the Flinders Ranges of South Australia, quartz grains are cemented together in a hard, pinkish rock. The rock, about 700 million years old, may have formed in deep water during brief storms. In 1946, at Ediacara, fossils of some of the oldest known multicelled animals were discovered in these rocks. Impressions of delicate fronds, ridged disks, and segmented circles are all that are left of this mysterious group of soft-bodied animals. Segmented *Spriggina* might have been an annelid worm, a type of trilobite, or even an anchored frond. Other fossils might be jellyfish or even filled-in burrows of animals.

IN THE BEGINNING
The work of paleontologists is to find fossils and piece together the story of ancient life on Earth. But we also want to know how life began, and even the earliest fossils do not answer that question. In 1953, U.S. chemist Stanley Miller showed how certain complex chemicals on which life depends – the building blocks of protein – could have first been produced on the Earth.

Powerful electric charge was fed into chamber

Spark reaction in gases produced amino acids

BRIGHT SPARK
Before blue-green algae produced oxygen-rich air, the Earth's atmosphere may have been an unhealthy mixture of gases including methane, ammonia, hydrogen, and water vapor. Dr. Miller's experiment showed that these gases can be triggered by a powerful electric spark to produce amino acids. Amino acids are the chemicals which make protein molecules, which in turn are the main substance of living things.

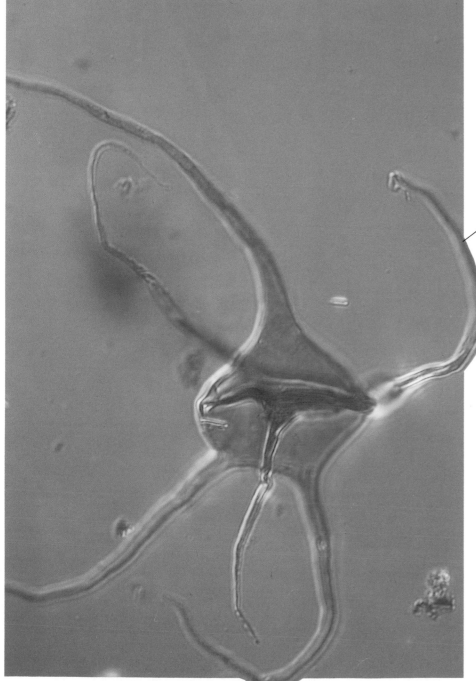

Spiny acritarch (Acanthomorph *acritarch*) *extracted from Silurian rocks*

An example of a carbonaceous chondrite meteorite, containing relatively high levels of carbon

METEORITE MOLECULES
Evidence that the molecules needed for life can form naturally comes from outer space. One meteorite, of a type called carbonaceous chondrites, fell near Murchison, Australia, in 1969. Scientists who studied it discovered amino acids just like those made artificially in Dr. Miller's experiments.

CONFUSED REMAINS
Understanding fossil life can be difficult. Sometimes this is reflected in the names given to fossils. Crushed and flattened, spiky or smooth, a group of microscopic spheres, ellipsoids, and three-sided pouches that first appeared about 1,800 million years ago are known as acritarchs, from the Greek for "of uncertain origin." They are some of the earliest known eukaryotes, and were sea-going microalgae.

Diversity of life

FOLLOWING THE STORY OF LIFE ON EARTH is rarely simple. Many animals have left no trace and the fossil record is far from complete. In the book of life, it is difficult to make sense of what is happening on one page when the pages on either side are missing. The fossils of the 530-million-year-old Burgess Shale in Canada's Rocky Mountains make a very unusual page in life's story. After more than 100 million years of blank spaces, the silvery traces preserved in thin sections of slate appear without much warning. Here, 8,000 ft (2400 m) above sea level, a slaty rock layer contains an incredible catalog of soft-bodied and thin-shelled sea animals that were very different from the creatures which left their mark at Ediacara (pp. 12–13) long before. Although many Burgess Shale fossils are similar to today's living animals, others have a unique body shape seen nowhere else before or since. It seems that 530 million years ago life may have been even more varied than today.

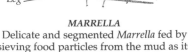

MARRELLA
Delicate and segmented *Marrella* fed by sieving food particles from the mud as its pairs of jointed legs carried it over the surface of the sea floor. Paired feathery gills took in oxygen from the water.

Head shield · *Antenna* · *Spine* · *Leg* · *Gill* · *Body segment*

Antenna · *Head shield* · *Body segment*

Rockies guide Tom Wilson

Prof. Walcott

DISCOVERER
Prof. Charles D. Walcott (1850–1927), Secretary of the Smithsonian Institution in Washington, D.C., discovered the Burgess Shale treasure in 1909. Over six field seasons, Walcott collected 65,000 specimens, but it is only in recent decades that paleontologists have come to appreciate the novelty of these fossils, which are mostly in the collections of the Smithsonian and the Geological Survey of Canada in Ottawa.

MARVELOUS MULTITUDE
Marrella is the most common animal fossilized in the Burgess Shale – more than 13,000 specimens have now been collected. Up to 0.75 in (2 cm) long, *Marrella* had a head shield from which two pairs of spines swept back over its body. The body had 24–26 segments, and each segment was biramous (two-branched). The upper branch had long gills, the lower a walking leg. Fossils of *Marrella* show the exceptional preservation of detail among the Burgess Shale animals. Swept up in an underwater landslip and quickly buried, the bodies were sealed before they could rot and fall apart.

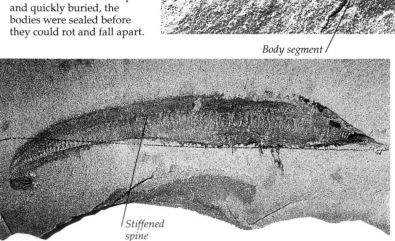

Stiffened spine

ODD ONE OUT
Among all the invertebrates (animals without backbones) from the Burgess Shale, such as worms, echinoderms, sponges, and arthropods, 1.5-in (4-cm) long *Pikaia* is the odd one out. It looks like a fish, but Walcott identified it as an annelid (segmented worm). Scientists now believe the ghostly *Pikaia* may be the first known chordate (an early vertebrate) and a forerunner of the vertebrate group that includes humans.

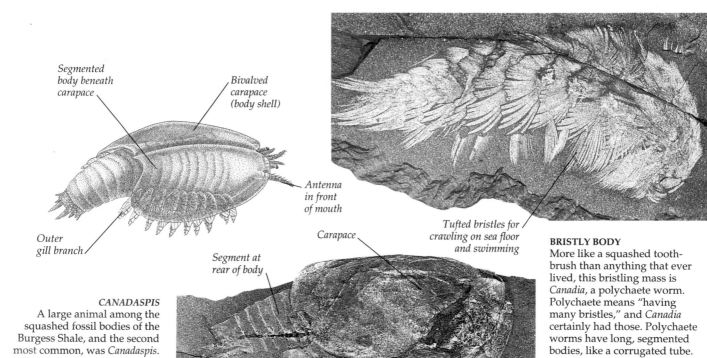

Segmented body beneath carapace

Bivalved carapace (body shell)

Antenna in front of mouth

Outer gill branch

CANADASPIS

A large animal among the squashed fossil bodies of the Burgess Shale, and the second most common, was *Canadaspis*. Carrying a bivalved carapace (body shell), 3-in (7.5-cm) long *Canadaspis* has been shown to be a distant relative of shrimps, prawns, and lobsters. Underneath the shell was a head with eyes, a body of 15 segments, and 10 pairs of legs and gills.

Segment at rear of body

Carapace

Tufted bristles for crawling on sea floor and swimming

Gill

Leg

BRISTLY BODY

More like a squashed tooth-brush than anything that ever lived, this bristling mass is *Canadia*, a polychaete worm. Polychaete means "having many bristles," and *Canadia* certainly had those. Polychaete worms have long, segmented bodies, like a corrugated tube. Each body ring has a pair of bristly tufts, which are used for crawling and swimming. Living polychaetes include the ragworm, which burrows in muddy sea shores.

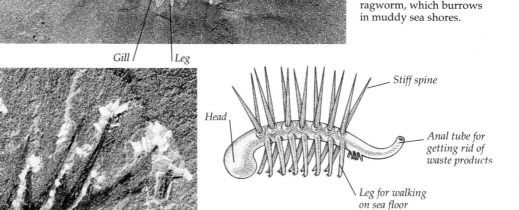

Head

Spine

Stiff spine

Head

Anal tube for getting rid of waste products

Leg for walking on sea floor

WHICH WAY UP?

New discoveries in China of fossils of *Hallucigenia* as old as those from the Burgess Shale suggest that the original scientific reconstruction of this animal was upside down, but the correct version is shown here. Seven pairs of stilts and long feeding tentacles turned out to be back spines and walking legs.

BAD DREAMS

Perhaps the most curious of all the creatures that lived on the Burgess sea floor was the stilted 1-in (2.5-cm) *Hallucigenia*. What kind of animal has seven pairs of spines, seven tentacles, and a tube-shaped body with a narrower tube at one end and a balloon at the other? *Hallucigenia* was so bizarre that scientists could not be sure which end was which or even which way was up! New discoveries in China have identified *Hallucigenia* as a caterpillar-shaped velvet worm. Another gap in the fossil record has been filled.

Leg

Anal tube

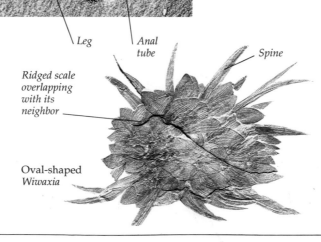

Spine

Ridged scale overlapping with its neighbor

Oval-shaped *Wiwaxia*

A FEATHERED CAP

Scales and spines are almost all there is to see of oval-shaped, 1.5-in (4-cm) *Wiwaxia*, an animal almost unknown anywhere but at Burgess Shale, but perhaps distantly related to mollusks. The ridged, flattened scales overlap, making a scaly cap. The spines stand up in two rows to defend the animal. A pair of tiny toothed bars were used for feeding. One other piece of evidence gives a fuller picture of *Wiwaxia*. Fossil brachiopods, attached to some scales, tell us that whatever *Wiwaxia* was, it crawled along the muddy sea floor and not under it, where brachio-pods would not have survived.

15

Hard covers

Model of *Cothurnocystis elizae*

Small slit for getting rid of "used" water

"Tail" (stem) for moving over sea floor

Dactylioceras commune Early Jurassic, U.K.

IN THE 110 MILLION YEARS between the creatures of Ediacara (pp. 12–13) and the diverse life of the Burgess Shale (pp. 14–15), life under the sea took a great evolutionary advance. Soft-bodied creatures evolved into animals with hard shells. Calcite (the calcium carbonate mineral in chalk and limestone) was one of the main materials for building shells. Animals grew calcite, forming a complete stony covering or patchwork of plates. Wrap-around shells transformed life in the oceans. Trilobites, free to swim or to crawl over the sea bed, flexed their exoskeletons (external skeletons) with internal muscles. In their fortress homes cemented to rocks, corals and brachiopods could grow and feed. Shells and exoskeletons let animals create their own living environments, safe from predators. Thanks to these hard bodies, millions of fascinating fossils are preserved in rocks all over the world.

CARVED HEAD
In the Middle Ages, ammonites – fossil shells – were thought to be snakes turned to stone; snakes' heads were carved on them.

Mouth at open end of "boot"

Boot-shaped head

ANIMAL ANCESTOR?
Like an odd-shaped fork, and small enough to hold in your hand, *Cothurnocystis elizae* (440 mya) was a strange but important animal. Some scientists believe that its ancestors were also the ancestors of all animals with backbones, including humans. Covered in hard, calcite plates, the boot-shaped head was pulled across the mud by its long stem. Water was drawn in the open end of the "boot" and filtered for food before being pushed out of the slits around the "toe."

Slit at "toe" end

Cothurnocystis elizae Late Ordovician Scotland

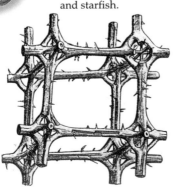

Tiny tube extended from plated arm to catch food

Crystalline plate covering body

Rings of calcite form a flexible stem

SEA LILY
Sometimes called sea lilies, crinoids such as this *Woodocrinus* were anchored in sea-bed gardens by long, jointed stems of calcite. Above the sea floor, food was caught by sticky tubes along the jointed waving arms and carried via grooves to a mouth on the plated cup. Crinoids, some of which still live today, first appeared about 500 mya. Although they look like plants, crinoids are close relatives of sea urchins and starfish.

Raphidonema faringdonense (sponge), Mid Cretaceous England

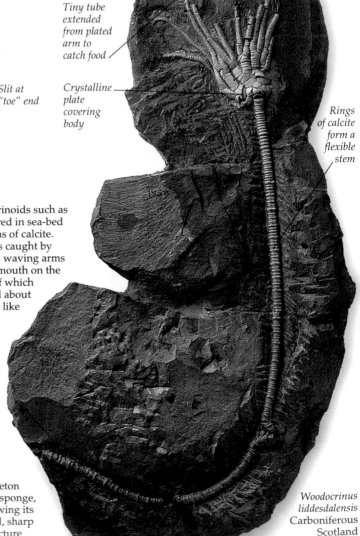

SPIKY SPONGES
Soft and luxurious, many sponges actually have a basketwork skeleton of tiny struts and spikes. This vase-shaped sponge skeleton is made of calcite. Others are made of silica, the same hard mineral which makes glass. These skeletons formed a cagelike framework that held the sponge body together and anchored it to the sea floor.

Skeleton of a sponge, showing its hard, sharp structure

Woodocrinus liddesdalensis Carboniferous Scotland

PROTECTIVE OVERCOAT

Trilobites dominated the earliest sea world of hard cases and survived until about 245 mya. Unlike bivalves and corals, which grow larger by adding material around the rims of their shells, trilobites had to cast off their "straitjackets." Exposed and in danger, their bodies grew and produced a new, larger cover to regain protection. This process happened several times. Many trilobites' cast-offs were fossilized, giving a false idea of trilobite numbers.

Large eye was composed of many lenses

When trilobite rolled up body for safety, each body segment hinged with its neighbor

Tail locked into groove under head for protection

Phacops africanus Devonian Western Sahara

Opening where tough stalk anchored brachiopod to sea bed

Large, curved eye gave excellent all-round vision

Head (cephalon) concealed stomach and mouth

Line of weakness in trilobite's skeleton allowed it to be discarded for a new, larger skeleton

Body segment

Tail is a single plate of fused segments

Phacops rana Devonian Ohio

Hinge opened and closed both parts of shell

FEEDING SHELLS

At first glance, brachiopods seem indistinguishable from the many bivalved sea shells found today. In fact, they belong to a very different animal group, which first appeared in the Early Cambrian. Brachiopods grow by adding crystals and organic material from the outer edges of the body to the surrounding edges of the shell. When the shell is open, tiny waving threads on the sticky lophophore (a ring of tentacles) draw in water while the lophophore filters food from the sea.

Spirifer Carboniferous Ireland

EMPIRE BUILDING

Fantastically shaped coral reefs are rich in sea life. In a modern coral reef, hard multistoried colonies of living coral build on top of cemented coral skeletons. The earliest known reefs were made by algae about 2 bya, but corals did not join the reef builders until the middle of the Ordovician period, about 470 mya. Today's coral reefs grow in shallow waters at temperatures above 64°F (18°C).

THE FIRST EYE

Trilobites were the first creatures to see. Their excellent vision detected movement and size in low levels of light in the water. Like other sea creatures with hard covers, they extracted from digested food and sea water the minerals needed for their shells. After death, the shells became part of the sediment on the sea floor. Fossil trilobites, with pairs of legs under the head, tail, and body segments, are often preserved without any trace of their delicate antennae or legs. Beside each leg was a feathery breathing gill, which may also have aided in swimming. Body segments were grown one at a time until the full adult number was reached.

Cup where polyp sat when it was alive

Ketophyllum Silurian Sweden

CORAL LIFE

This many-branched fossil coral was once a colony of fleshy animals. A soft-bodied coral animal (polyp), sat in each hollow cup at the end of a branch. Feathery tentacles ringed its upward-facing mouth, swaying in the water currents. As the polyp grew, it added more calcium carbonate to the skeleton beneath, increasing the height and width of its stony tower. Deep folds in the animal's underside helped form radiating ribs that supported the growing coral structure.

The inside story

WITHIN 60 MILLION YEARS of Burgess Shale life, one group of animals, the vertebrates, escaped the restrictions of living in shells. They developed an internal bony skeleton that anchored muscles and supported internal organs. Bones are made of hard mineral crystals plus fibrous protein. Cells inside bones are fed by blood vessels, keeping the bones nourished. Bones, teeth, and scales are tough and preserve well as fossils. The first vertebrates were jawless fishes. They appeared 470 million years ago. Some, like cephalaspids and placoderms, carried a heavy outer armor that restricted them to living on sea beds. Later fish types had less bony covering on their heads and had toothed, gaping jaws. The success of advanced bony fishes (the teleosts) is seen in the vast numbers of these mobile fishes living today in rivers, lakes, and seas.

FROZEN FISH
Louis Agassiz (1807–1873), the Swiss-American naturalist, was a great student of fishes, and the discoverer of the movement of glaciers.

Head

Tail

MYSTERIOUS TEETH
For over 100 years, scientists debated whether the tiny, toothlike spikes of conodonts were remains of mollusks, fishes, worms, or plants. The mystery was solved in Scotland when a set of Early Carboniferous conodont fossils were found in the fossilized remains of a long, soft, eel-like body – possibly a primitive swimming chordate (early vertebrate).

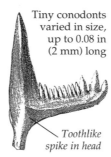

Tiny conodonts varied in size, up to 0.08 in (2 mm) long

Toothlike spike in head

HEAVY ARMOR
One of the most armored of placoderm fishes was the 370-million-year-old, 5-in (13-cm) *Pterichthyodes*. It had no inner bone skeleton. A shell of bony plates covered both the head and body. Even the pectoral fins were enclosed in a bony casing and would not have been of much use for swimming.

Spiny pectoral fin used for punting along the muddy sea bottom

Pterichthyodes milleri
Mid Devonian
Scotland

Homocercal (symmetrical) two-part caudal (tail) fin

Birkenia elegans
Silurian, Scotland

Dorsal spine

Lepidotes elvensis
Early Jurassic
Germany

SMALL SCALE
Distinctive 3-in (7-cm) long *Birkenia*, with its heterocercal (unequally divided), down-turned tail, was a freshwater jawless fish that lived about 425 mya. Overlapping bony scales covered the body, and several hooked spines protruded from the dorsal (top) surface. Small plates formed a mosaic on top of the head, while gill openings formed a diagonal dotted line behind.

SENSITIVE SUCKER
The first fishes were jawless, sucking food and water through their mouths. *Cephalaspis* had a bony shield covering its jawless head, as well as a pair of pectoral fins protected by swept-back spines. Two eyes and a single nostril perched on the crest of the arched head shield, which had three sensitive scale-covered patches connected to the brain. *Cephalaspis* and *Birkenia* were ostracoderms, fishes with "bony skins."

Cephalaspis pagei
Devonian, Scotland

Eye

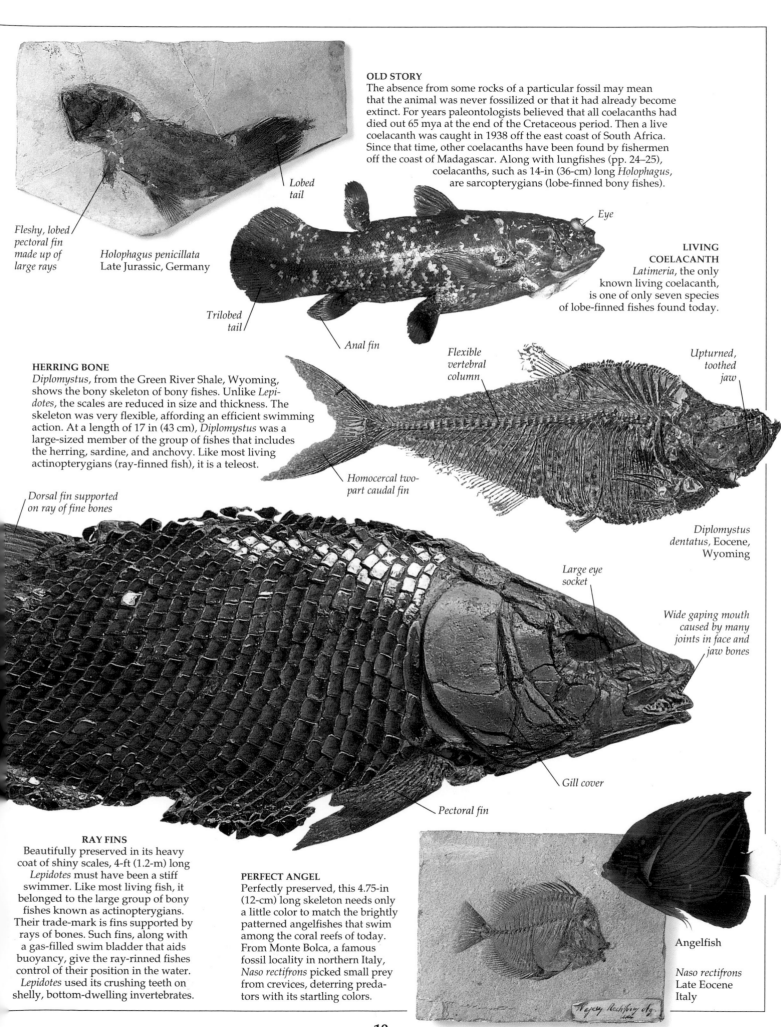

OLD STORY
The absence from some rocks of a particular fossil may mean that the animal was never fossilized or that it had already become extinct. For years paleontologists believed that all coelacanths had died out 65 mya at the end of the Cretaceous period. Then a live coelacanth was caught in 1938 off the east coast of South Africa. Since that time, other coelacanths have been found by fishermen off the coast of Madagascar. Along with lungfishes (pp. 24–25), coelacanths, such as 14-in (36-cm) long *Holophagus*, are sarcopterygians (lobe-finned bony fishes).

Lobed tail

Eye

LIVING COELACANTH
Latimeria, the only known living coelacanth, is one of only seven species of lobe-finned fishes found today.

Fleshy, lobed pectoral fin made up of large rays

Holophagus penicillata
Late Jurassic, Germany

Trilobed tail

Anal fin

Flexible vertebral column

Upturned, toothed jaw

HERRING BONE
Diplomystus, from the Green River Shale, Wyoming, shows the bony skeleton of bony fishes. Unlike *Lepidotes*, the scales are reduced in size and thickness. The skeleton was very flexible, affording an efficient swimming action. At a length of 17 in (43 cm), *Diplomystus* was a large-sized member of the group of fishes that includes the herring, sardine, and anchovy. Like most living actinopterygians (ray-finned fish), it is a teleost.

Homocercal two-part caudal fin

Diplomystus dentatus, Eocene, Wyoming

Dorsal fin supported on ray of fine bones

Large eye socket

Wide gaping mouth caused by many joints in face and jaw bones

Gill cover

Pectoral fin

RAY FINS
Beautifully preserved in its heavy coat of shiny scales, 4-ft (1.2-m) long *Lepidotes* must have been a stiff swimmer. Like most living fish, it belonged to the large group of bony fishes known as actinopterygians. Their trade-mark is fins supported by rays of bones. Such fins, along with a gas-filled swim bladder that aids buoyancy, give the ray-rinned fishes control of their position in the water. *Lepidotes* used its crushing teeth on shelly, bottom-dwelling invertebrates.

PERFECT ANGEL
Perfectly preserved, this 4.75-in (12-cm) long skeleton needs only a little color to match the brightly patterned angelfishes that swim among the coral reefs of today. From Monte Bolca, a famous fossil locality in northern Italy, *Naso rectifrons* picked small prey from crevices, deterring predators with its startling colors.

Angelfish

Naso rectifrons
Late Eocene
Italy

19

Sea monsters

TALES OF SEA MONSTERS have been told since people first set out to explore the world's great oceans and lakes. These myths still exist, from mermaids and Moby Dick, the legendary whale, to Scotland's Loch Ness monster. In the history of the world's seas, there has been no shortage of monstrous creatures. Many of the early forms of life were confined to living on the sea bed. Others actively hunted their neighbors where, in this competitive world, larger size was one good means of survival. Giant predators have appeared in many groups of animals. Among arthropods, the scorpion-like eurypterid, *Pterygotus*, reached the monstrous size of 6 ft 6 in (2 m) and hunted fish. Many other enormous fishes emerged, such as *Dunkleosteus* and later *Xiphactinus*, but it was the emergence of biting jaws among the fishes that helped them dominate other life forms in the seas, lakes, and rivers. Even the rise of ferocious marine reptiles (pp. 38–39) failed to overcome the continued success of fish.

SILURIAN SEASCAPE
Predatory eurypterids (aquatic, scorpion-like arthropods) dominated life in the sea during the Silurian period (439–408 mya). While most fishes were still without jaws and scouring the sea bed for food, eurypterids were among the most active hunters. Some later specimens were 6 ft 6 in (2 m) long.

Row of teeth rimmed bulldog-shaped jaw

WHAT A CATCH!
Of today's 21,700 fish species, almost 20,000 are teleosts (bony fish). No living bony fish surpasses *Xiphactinus* in size. When alive, this 13-ft 9 in (4.2-m) long specimen may have weighed as much as 1,650 lb (750 kg). *Xiphactinus* lived about 80 mya in the Cretaceous seas of Europe, Australia, and North America. The large conical teeth along the margins of its jaws were probably the last things its captured prey ever saw.

Jointed last leg with oar-shaped paddle

Eurypterus lacustris
Silurian, New York

Small chelicera (jointed pincer)

Operculum (gill cover)

Pectoral fin

PINCERS AND PADDLES
A distant relative of today's scorpions and spiders, eurypterids first appeared about 470 mya. They had the segmented body typical of arthropods, such as trilobites (pp. 16–17), and were equipped with two chelicerae (jointed pincers) for biting prey. This eurypterid specimen (shown from the underside) had six pairs of appendages: the pair of pincers in front of the mouth for catching food, four pairs of jointed legs for walking on the sea bed, and a pair of flat-ended paddles for swimming in the sea.

Fossilized Gillicus *facing tail of* Xiphactinus

Dorsal fin

Caudal fin

Pectoral fin

Pelvic fin

Anal fin

FISH FOOD
Paleontologists have no difficulty in finding out what *Xiphactinus* preyed on in the Cretaceous sea. Several complete fossil skeletons have been discovered with the complete fossil of another fish, *Gillicus*, lying inside the rib cage. *Gillicus* was no small fry itself at 6 ft (1.8 m) long. Some scientists have suggested that these "stuffed" *Xiphactinus* died of gluttony, unable to digest the whole fish they had swallowed!

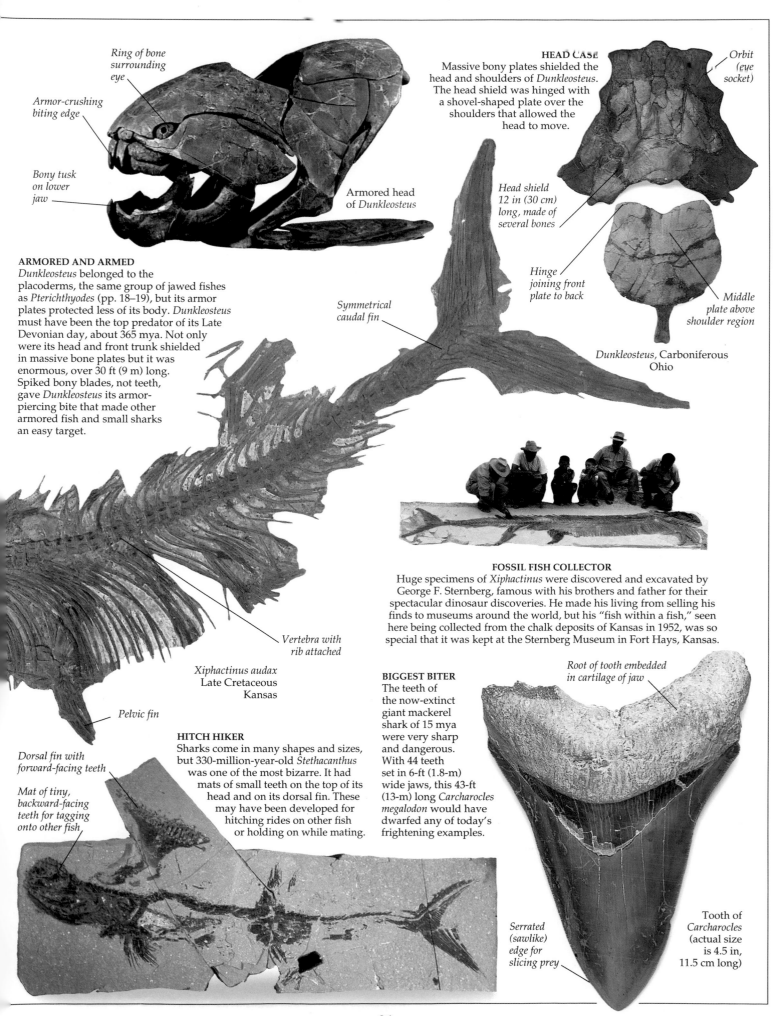

Ring of bone surrounding eye

Armor-crushing biting edge

Bony tusk on lower jaw

Armored head of *Dunkleosteus*

HEAD CASE
Massive bony plates shielded the head and shoulders of *Dunkleosteus*. The head shield was hinged with a shovel-shaped plate over the shoulders that allowed the head to move.

Orbit (eye socket)

Head shield 12 in (30 cm) long, made of several bones

Hinge joining front plate to back

Middle plate above shoulder region

Dunkleosteus, Carboniferous Ohio

ARMORED AND ARMED
Dunkleosteus belonged to the placoderms, the same group of jawed fishes as *Pterichthyodes* (pp. 18–19), but its armor plates protected less of its body. *Dunkleosteus* must have been the top predator of its Late Devonian day, about 365 mya. Not only were its head and front trunk shielded in massive bone plates but it was enormous, over 30 ft (9 m) long. Spiked bony blades, not teeth, gave *Dunkleosteus* its armor-piercing bite that made other armored fish and small sharks an easy target.

Symmetrical caudal fin

Vertebra with rib attached

Xiphactinus audax
Late Cretaceous
Kansas

Pelvic fin

FOSSIL FISH COLLECTOR
Huge specimens of *Xiphactinus* were discovered and excavated by George F. Sternberg, famous with his brothers and father for their spectacular dinosaur discoveries. He made his living from selling his finds to museums around the world, but his "fish within a fish," seen here being collected from the chalk deposits of Kansas in 1952, was so special that it was kept at the Sternberg Museum in Fort Hays, Kansas.

HITCH HIKER
Sharks come in many shapes and sizes, but 330-million-year-old *Stethacanthus* was one of the most bizarre. It had mats of small teeth on the top of its head and on its dorsal fin. These may have been developed for hitching rides on other fish or holding on while mating.

Dorsal fin with forward-facing teeth

Mat of tiny, backward-facing teeth for tagging onto other fish

BIGGEST BITER
The teeth of the now-extinct giant mackerel shark of 15 mya were very sharp and dangerous. With 44 teeth set in 6-ft (1.8-m) wide jaws, this 43-ft (13-m) long *Carcharocles megalodon* would have dwarfed any of today's frightening examples.

Root of tooth embedded in cartilage of jaw

Serrated (sawlike) edge for slicing prey

Tooth of *Carcharocles* (actual size is 4.5 in, 11.5 cm long)

Taking root on land

LIVING IN WATER HAS ADVANTAGES. There is no risk of drying out, and food streams by. Organisms do not even have to support their full weight – the water does it for them. When plants moved out of water onto dry land they needed new ways to survive. The oldest-known land plant, *Cooksonia*, lived about 400 million years ago. Long, corrugated tubes of cells (tracheid tubes) kept plants upright and also provided a system of channels for carrying water up the stem. Roots absorbed water from the soil. These new "vascular" plants developed cuticle, a waterproof coating that kept moisture in, with pores that allowed plants to "breathe." Wherever plants appeared, arthropods followed. Plants offered arthropods a food supply and a new habitat. Close relationships among millions of insects and plant species continue today.

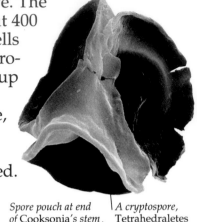

Spore pouch at end of Cooksonia's *stem* | *A cryptospore,* Tetrahedraletes

SILURIAN GREENERY
Vascular plants appeared near the end of the Silurian period. Many of these plants were small, like *Cooksonia*. Some lived in damp, swampy habitats where green algae were already present. Arthropods browsed through dead vegetation, processing it and improving the quality of the soil.

Sporangium at end of single stem of modern liverwort

Single sporangium on tip of stem

Disk-shaped sporangium on side of stem

Leaflike scale of Asteroxylon

Early reconstruction of Aglaophyton's *simple root system with three-layered stems*

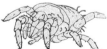

EARLY PLANTS
Primitive land-plants reproduced by releasing spores into the wind. *Cooksonia* (like the modern liverwort) carried spores in 0.04-in (1-mm) diameter pouches (sporangia) on the end of flattened, branching stems. Cryptospores, such as *Tetrahedraletes*, are known from the Devonian period and may belong to the family of bryophytes, plants which do not have a vascular system that carries water.

Modern springtail similar to the first known insect, *Rhyniella*, found at Rhynie

Protacarus, a predatory mite from Rhynie (Devonian period)

Fossilized plants are contained in this piece of Rhynie Chert, a silica-rich rock from Scotland

Hot waters
Rhynie plants were preserved by silica-rich water from volcanic hot springs. A thick gel of silica (a quartz-like mineral) preserved the plants and many stems, spores, and leaves. Even the remains of spiders, mites, and other arthropods have been preserved. *Asteroxylon*, 20-in (50-cm) tall, carried large, flat spore pouches (sporangia) among its scaly leaves. *Aglaophyton major* had 7-in (18-cm), smooth stems and combined features of bryophytes and vascular plants.

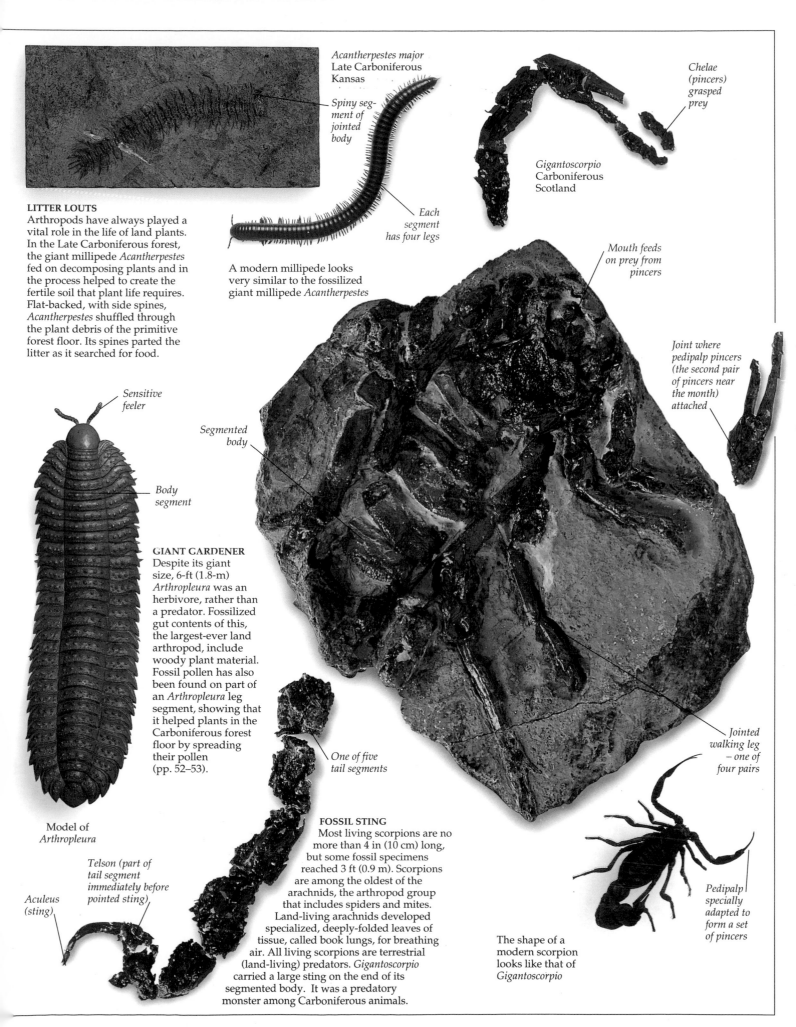

Acantherpestes major
Late Carboniferous
Kansas

Spiny seg-
ment of
jointed
body

*Chelae
(pincers)
grasped
prey*

Gigantoscorpio
Carboniferous
Scotland

Each
segment
has four legs

LITTER LOUTS
Arthropods have always played a
vital role in the life of land plants.
In the Late Carboniferous forest,
the giant millipede *Acantherpestes*
fed on decomposing plants and in
the process helped to create the
fertile soil that plant life requires.
Flat-backed, with side spines,
Acantherpestes shuffled through
the plant debris of the primitive
forest floor. Its spines parted the
litter as it searched for food.

A modern millipede looks
very similar to the fossilized
giant millipede *Acantherpestes*

*Mouth feeds
on prey from
pincers*

*Sensitive
feeler*

*Segmented
body*

*Joint where
pedipalp pincers
(the second pair
of pincers near
the month)
attached*

*Body
segment*

GIANT GARDENER
Despite its giant
size, 6-ft (1.8-m)
Arthropleura was an
herbivore, rather than
a predator. Fossilized
gut contents of this,
the largest-ever land
arthropod, include
woody plant material.
Fossil pollen has also
been found on part of
an *Arthropleura* leg
segment, showing that
it helped plants in the
Carboniferous forest
floor by spreading
their pollen
(pp. 52–53).

*Jointed
walking leg
– one of
four pairs*

*One of five
tail segments*

Model of
Arthropleura

FOSSIL STING
Most living scorpions are no
more than 4 in (10 cm) long,
but some fossil specimens
reached 3 ft (0.9 m). Scorpions
are among the oldest of the
arachnids, the arthropod group
that includes spiders and mites.
Land-living arachnids developed
specialized, deeply-folded leaves of
tissue, called book lungs, for breathing
air. All living scorpions are terrestrial
(land-living) predators. *Gigantoscorpio*
carried a large sting on the end of its
segmented body. It was a predatory
monster among Carboniferous animals.

*Telson (part of
tail segment
immediately before
pointed sting)*

*Aculeus
(sting)*

*Pedipalp
specially
adapted to
form a set
of pincers*

The shape of a
modern scorpion
looks like that of
Gigantoscorpio

First four feet

Stepping from their watery habitat onto dry land, the earliest amphibians had many fishlike features, including a wide, fishy tail. Clumsy in appearance, with short, squat limbs, these first tetrapods – four-footed animals – had no competition in their new land life. In water, their ancestors had been just "small fish" in a big pool, but on land tetrapods entered a new world. To live on land, tetrapods had to be able to breathe air and have a skeleton and muscles strong enough to support their weight out of water. The newest evidence shows that legs, which seem essential on dry land, were developed first in aquatic ancestors. Amphibians are not free of their fishy past. Their skin is not equipped for a dry world, and their eggs cannot survive out of water. Today's moist-skinned frogs, toads, and newts still depend on wet habitats for survival.

DEVONIAN SCENE
The Devonian period (408–362 mya), when vertebrates first walked on land, was marked by arid climates in many parts of the world. Lobe-finned fishes, the ancestors of tetrapods, were common in shallow lakes and rivers.

WALK ON WATER
Paleontologists have long identified an extinct group of lobe-finned fishes, rhipidistians, as ancestors of early tetrapods. Today, some scientists believe the Dipnoi (living and fossil lungfishes) are more closely related to tetrapods. For example, if its lake becomes shallow and stagnant, *Neoceratodus* pushes along the lake floor on its fins and breathes air.

Dipterus valenciennesi (a fossil lungfish), Devonian, Scotland

Lungfish *Neoceratodus forsteri* lives only in Australia

Roof of skull made of four flexible sections

Model of *Eusthenopteron foordi*

Fossil of *Eusthenopteron foordi* Late Devonian Canada

Muscly pectoral fin used for steering

Three-lobed tail propelled fish through water

Pelvic fin

OUT ON A LIMB
Fossils of *Eusthenopteron*, a 30-in (75-cm) long rhipidistian fish, come from Late Devonian rocks of Scotland and Canada. Many features of its skeleton are similar to those of the earliest amphibians. The pattern of skull bones is like that of amphibians, while the bones within the pectoral and pelvic fins have been directly compared to the limb bones of tetrapods. But the image of fishes dragging their fins overland during droughts in search of pools of water is far-fetched.

Shoulder girdle

Pectoral fin ray

Strong back

Fish-finned tail

Fishlike scales

Strengthened hip girdle

This model of Ichthyostega *shows five-toed, webbed, paddle feet, but it is now known that each hind foot had seven toes*

Spiracle (vent) drew in water

Eye socket

Fossil skull of *Acanthostega* Late Devonian Greenland

Fishlike tail fin enhanced swimming

Reconstruction of *Acanthostega*

Pocket encloses nostril for drawing in air

Vertebral column supported strong muscles for swimming

Operculum (gill cover)

Tail

Row of tiny teeth

FISH OUT OF WATER

Discovered in 1952, the fossils of *Acanthostega* are the earliest remains of tetrapods ever found. Bearing fishlike gills and a tail with a fin, 3-ft (1 m) long *Acanthostega* lived more in water than it did on land. The fossils are evidence that feet first developed in aquatic animals, rather than animals that were already land-based. *Acanthostega* probably had lungs, too, filling them with air pumped through its mouth as it crawled onto lake shores or river banks.

Gill chamber containing gill bars

Recently discovered specimen of *Acanthostega*

Eye socket

Toes embedded in rock

Part of pelvis

Femur (thigh bone)

Fibula (calf bone)

Strong vertebral column supported body weight

Wide, U-shaped shoulder girdle, separate from back of skull

Broad pelvic girdle carried muscles that lifted body onto legs

QUITE A HANDFUL

Five fingers and toes may seem normal, but *Ichthyostega* is now known to have had seven toes on its hind feet rather than the five first suggested. *Acanthostega* had even more toes, with eight on each of its front feet.

Tibia (shin bone)

Clump of three tiny toes

Flat, paddle-shaped hind limb of *Ichthyostega* Late Devonian, Greenland

Skeleton of *Eryops* Permian, Texas

BONY BULK

Broad-shouldered, 6-ft (2-m) long *Eryops* was an awkward mover on land. Its wide, long, flat skull had sharp, spiky teeth for attacking prey. Two hundred million years after the tetrapods first walked on land, many different kinds of amphibians lived a semi-aquatic life. Even for large *Eryops*, however, competition was already present in up-and-coming reptiles.

Model of *Ichthyostega* Late Devonian Greenland

FIN AND FEET

Ichthyostega, one of the first tetrapods, was 3 ft (1 m) long and lived about 360 mya. Fossils of *Ichthyostega* have been found only in Greenland, where *Acanthostega* was also found. This model shows the generally fishy outline of *Ichthyostega* even though it was built before the latest fossil finds. A fish-finned tail would have propelled it through the water and, if it walked on land, the same sinuous body movement might have helped it sidle forward to catch prey in its fanglike teeth.

Strong shoulder girdle

Front foot has five toes

Forest swamps

THE SEEDS OF INDUSTRIAL DEVELOPMENT, with its early reliance on coal, were sown about 330 million years ago. In the Carboniferous period, a warm, humid climate prevailed over much of the northern and southern continents, and vast swamps supported dense forests of giant treelike plants. Great thicknesses of dead vegetation were laid down. Over millions of years they were converted to carbon-rich coal. The Carboniferous coal swamps were filled with 66-ft (20-m) tall horsetails, huge ferns, and giant club mosses such as *Lepidodendron*. Each fossilized part of this tree has its own name. The fossils can be combined to rebuild this giant of the forest swamps.

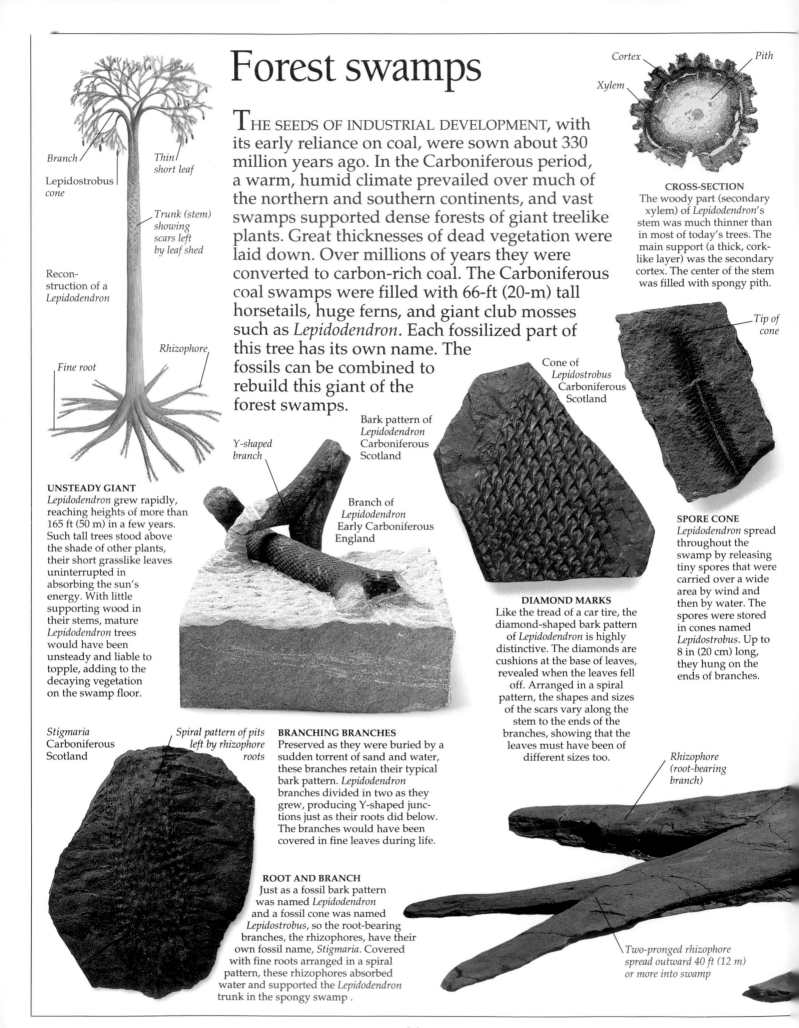

Branch

Thin short leaf

Lepidostrobus cone

Trunk (stem) showing scars left by leaf shed

Recon-struction of a *Lepidodendron*

Rhizophore

Fine root

Cortex

Pith

Xylem

CROSS-SECTION
The woody part (secondary xylem) of *Lepidodendron's* stem was much thinner than in most of today's trees. The main support (a thick, cork-like layer) was the secondary cortex. The center of the stem was filled with spongy pith.

Tip of cone

Cone of *Lepidostrobus* Carboniferous Scotland

Bark pattern of *Lepidodendron* Carboniferous Scotland

Y-shaped branch

Branch of *Lepidodendron* Early Carboniferous England

UNSTEADY GIANT
Lepidodendron grew rapidly, reaching heights of more than 165 ft (50 m) in a few years. Such tall trees stood above the shade of other plants, their short grasslike leaves uninterrupted in absorbing the sun's energy. With little supporting wood in their stems, mature *Lepidodendron* trees would have been unsteady and liable to topple, adding to the decaying vegetation on the swamp floor.

SPORE CONE
Lepidodendron spread throughout the swamp by releasing tiny spores that were carried over a wide area by wind and then by water. The spores were stored in cones named *Lepidostrobus*. Up to 8 in (20 cm) long, they hung on the ends of branches.

DIAMOND MARKS
Like the tread of a car tire, the diamond-shaped bark pattern of *Lepidodendron* is highly distinctive. The diamonds are cushions at the base of leaves, revealed when the leaves fell off. Arranged in a spiral pattern, the shapes and sizes of the scars vary along the stem to the ends of the branches, showing that the leaves must have been of different sizes too.

Stigmaria Carboniferous Scotland

Spiral pattern of pits left by rhizophore roots

BRANCHING BRANCHES
Preserved as they were buried by a sudden torrent of sand and water, these branches retain their typical bark pattern. *Lepidodendron* branches divided in two as they grew, producing Y-shaped junc-tions just as their roots did below. The branches would have been covered in fine leaves during life.

Rhizophore (root-bearing branch)

ROOT AND BRANCH
Just as a fossil bark pattern was named *Lepidodendron* and a fossil cone was named *Lepidostrobus*, so the root-bearing branches, the rhizophores, have their own fossil name, *Stigmaria*. Covered with fine roots arranged in a spiral pattern, these rhizophores absorbed water and supported the *Lepidodendron* trunk in the spongy swamp .

Two-pronged rhizophore spread outward 40 ft (12 m) or more into swamp

How coal is formed

Coal deposits have built up during many geological periods from the Devonian onwards, but those of the Carboniferous are the best known. When swamp plants died and fell into water-logged soil, the water prevented microorganisms from fully breaking down the plant tissue. Instead it was left as a light, spongy peat, preserving many of the plants and animals found today as fossils. New plants grew on the wet peat surface, piling up more dead plant material when they toppled. Periodic flooding covered this vegetation with layers of sand and mud. Over millions of years, the peat was compressed by rock layers into thick seams of coal.

FOSSIL FUEL
Coal has been the power behind the steam-powered engines and the growth of the chemical and transport industries during the Industrial Revolution of the 1800s. Vast supplies of this high-energy fuel and raw material for products lie under-ground in many countries – a huge economic resource.

Fossil fuel in the form of a lump of coal

BURIED TREE
Sigillaria was another large tree from the Carboniferous coal swamps. Over 3 ft 3 in (1 m) in diameter, *Sigillaria* had a trunk more than 100 ft (30 m) high and a fringe of long, grasslike leaves hanging from one or two branches. Upright trunk casts found in coal seams show how fast some swamps were flooded with sediments.

SWAMPS – DEAD OR ALIVE
Carboniferous coal-forming swamps, packed with live and dead vegetation and a variety of animal life, may have looked like the Everglades in Florida. Arthropods fed on rotting vegetation and on each other, while the large predators – the amphibians – were swimming in pools and preying on fish.

Cast stump, preserved by in-filling muds, is buried in surrounding coal and rock

Natural cast bears marks from inside of stump

Rhizophore for drawing water up woody stem

Rhizophore branch carried roots similar to leaves on upper branch

Rhizophore for securely anchoring Lepidodendron in swampy ground

FOSSIL GROVE
Lepidodendron trunks and root still standing where they once grew are not rare. The most famous are a group of eleven in a park in Glasgow, Scotland. These fossils preserve none of the original plant material; they are casts of the inside of the trunk and roots. The trees had broken and their stumps had rotted, leaving only the toughest outer layer. A sand-loaded flood poured enormous quantities of fine sediment into the stand of hollow stumps, filling them down to the roots. The surrounding rock has been removed, leaving the casts standing free.

Fossilized tree stumps of *Lepidodendron*
Carboniferous, Scotland

Reptiles reign

EVEN IF AMPHIBIANS can survive on dry land, they must return to water to lay their soft eggs. Amniotes are animals that freed themselves of this need by producing an egg that enclosed the developing embryo in its own wet world, protecting it with a tough, waterproof shell. Reptiles were the first amniotes, followed by birds and mammals. The first reptiles appeared in the Carboniferous period and may have looked little different from their amphibian ancestors. Unfortunately, there are no fossils of internal egg structures, and paleontologists must distinguish between amphibians and early reptiles by skeletal differences.

HOT STUFF
Reptiles adopted many shapes. In the Early Permian period, *Dimetrodon* (a meat-eating pelycosaur) grew to 10 ft (3 m) long and carried a huge fan of skin along its back as a giant radiator and heat absorber.

SETTING SAIL
A plant-eating pelycosaur (sailed lizard), *Edaphosaurus* had a skin-covered sail supported on long vertebral spines with cross pieces, like a ship's mast.

Pelvis, legs, and five-toed feet were separated from trunk

Individual spine in backbone

One of the spines that supported the sail of *Edaphosaurus*, Early Permian, U.S.

Dry, scaly skin

Flattened skull

Backbone with ribs attached

REPTILE FIRST
From a small quarry in central Scotland, this distorted 8-in (20-cm) long fossil skeleton (nicknamed "Lizzie") is one of the most important fossils ever found. Sandwiched between fine rock layers for 338 million years, *Westlothiana lizziae* is the oldest reptile and amniote known.

Very long body

Model of *Westlothiana lizziae* (viewed from above), based on scientific research and artist's impression

UNDERSTANDING THE EVIDENCE
Westlothiana was discovered in Early Carboniferous rocks that formed in a lake fed by hot, volcanic spring water. Four-legged, as well as legless, snake-like amphibian skeletons were also found, along with spiders, scorpions, and millipedes. Tree-sized seed ferns grew nearby. *Westlothiana* was about 1 ft (30 cm) long, with short legs. Some, but not all, features of the skull and limb bones match those found in later reptiles. It is not surprising that *Westlothiana*, so similar to its amphibian ancestors, is not easily distinguished from them.

Long, lizard-like tail

Tail made up large part of whole body length

Scales on skin helped keep Westlothiana waterproof

Model of *Westlothiana lizziae*, the earliest known reptile Early Carboniferous Scotland

Each foot had five toes

Skull, vertebrae and ribs, and underside of hind limbs of *Procolophon* Early Triassic South Africa

Backbone

Wide back teeth

Forelimb

Left leg

Right foot

COVERED SKULL

The anapsids, like large-eyed *Procolophon*, were primitive reptiles that had a complete cover of bone behind the eye sockets, with no skull openings, while more advanced reptiles had one or two skull openings.

BACK IN THE SWIM

Although reptiles developed as land-based life, they were able to adapt to life at sea. Mesosaurs, such as *Stereosternum*, were the first reptiles to return to water, propelled by rear-foot paddles.

Fossil skeletons of *Stereosternum* Early Permian Brazil

Long snout lined with spiky, water-straining teeth

Rear feet and long tail propelled mesosaurs through water

Pineal opening (third eye)

Stout limb to support bulky body

Skull of *Rhynchosaurus* Late Triassic England

PLANT PLOW

Rhynchosaurus's snout hooked over its lower jaw and ended in a sharp, digging spike, for tearing plants from the ground and raking them between its jaws. Tightly packed teeth in the lower jaw cut against a tooth-lined groove in the upper. Large muscles closed the jaws tight to cut through plants. The teeth of adults wore down, to leave a sharp-edged jaw bone.

Belodon, a crocodile-like Triassic reptile

Fossil skeleton of *Deltavjatia vjatkensis* Permian, Russia

Skull is small in proportion to body

LARGE-SCALE REPTILES

In the Late Permian period, a group of anapsid reptiles, the pareiasaurs, appeared as great stocky creatures up to 10 ft (3 m) long. Pareiasaurs, such as *Deltavjatia* from Russia, had stout vertical limbs to support their bulk. Their skulls were large and wide, with bony spikes along the sides of their skulls, as well as bony lumps embedded in their skin. Pareiasaurs were plant eaters, and had teeth like those of living lizards.

Shape of Westlothiana lizziae looks very similar to today's four-footed reptiles

Short, small, well-developed leg

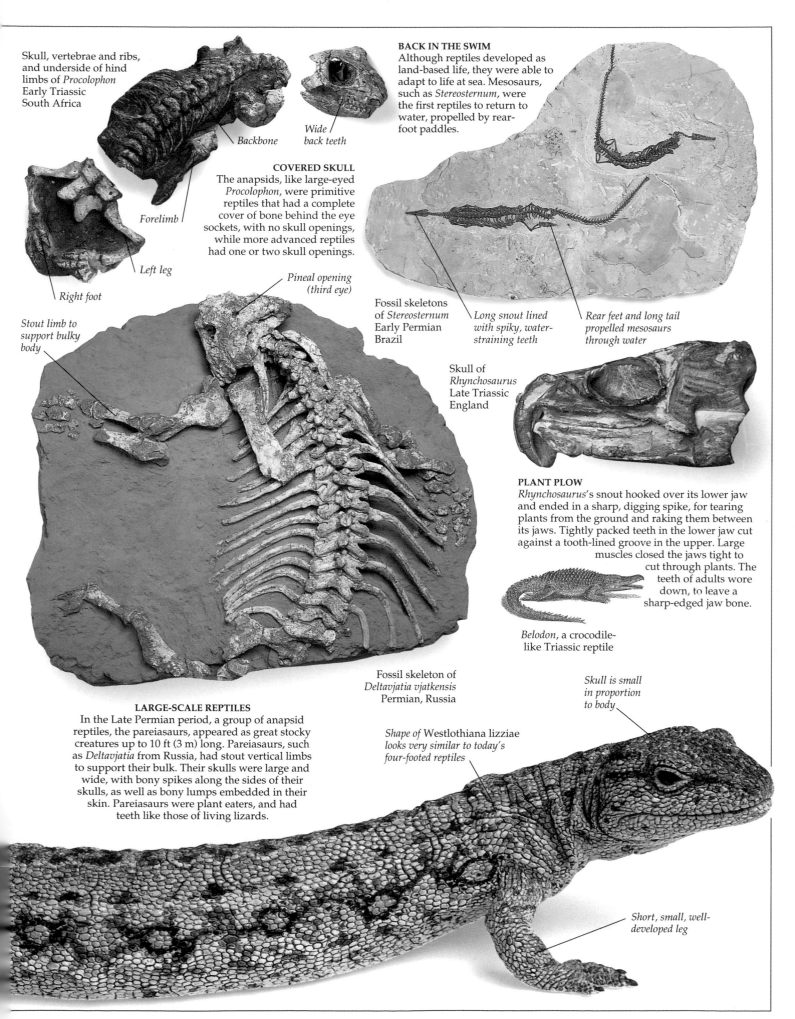

Reptiles like mammals

TRACING THE FOSSIL STORY OF LIFE is like connecting the dots. If the dots are far enough apart, they seem to mark a straight line. The development to mammals from reptiles was far from straight, as one after another reptile group evolved and then disappeared. Among all the reptile groups, only cynodonts gave rise to mammals. They belonged to a group of mammal-like reptiles called the synapsids. This group includes pelycosaurs, like the meat-eating *Dimetrodon*, and the plant-eating dicynodonts. Synapsids were distinguished from other reptiles by the single skull opening behind the eye sockets.

Skull of *Lystrosaurus murrayi*, Early Triassic South Africa

Eye socket

Hinge allowing jaw to slide back and forth

Canine-like tusk

Skull of *Dicynodon leoniceps* viewed from above, Late Permian, Tanzania

Forward facing eye socket

SLICING BEAKS
Toothless except for its canine tusks, *Lystrosaurus* was once the most common reptile in the Early Triassic world. A dicynodont, it had a horny beak covering the sharp cutting-edges of its short snout. Dicynodonts were herbivores, with the jaw hinged so that as the mouth opened and closed, the jaw's slicing edges slid forward and backward against each other, cutting up roots and stems.

MUSCLED SKULL
Dicynodon was a common dicynodont during the Late Permian period. From above, the 15-in (38-cm) long skull is wide, with great cheek-bones fanning out from the eye sockets. The huge spaces inside the cheeks were for powerful muscles that operated the slicing jaws. A narrow skull between the muscle spaces rose in a midline crest, increasing the area for muscle anchorage. Dicynodonts were the first large plant eaters. They lasted about 40 million years.

Wide cheek-bone around space once occupied by jaw muscles

Thick bone on roof

Skull of *Moschops* Late Permian South Africa

Beak covered short, steep snout

Maxilla (upper jawbone)

Dentary (lower jawbone)

Large neck vertebra

Eye socket

Tusk canine tooth, used for digging up plants and for defense

BONE HEAD
Moschops, a 16-ft 6 in (5-m) long plant eater, was one of the largest of the mammal-like reptiles. *Moschops* belonged to the dinocephalians, a group of primitive mammal-like reptiles (synapsids) from the Late Permian period of Russia and South Africa. Broad-spread legs, high at the front, carried *Moschops*'s great weight. Its skull roof was covered with 4-in (10-cm) thick bone, which together with strong neck vertebrae suggest that *Moschops* may have taken part in trials of strength, pushing at competitors with its head.

Reconstruction of primitive mammal-like reptile, *Moschops*

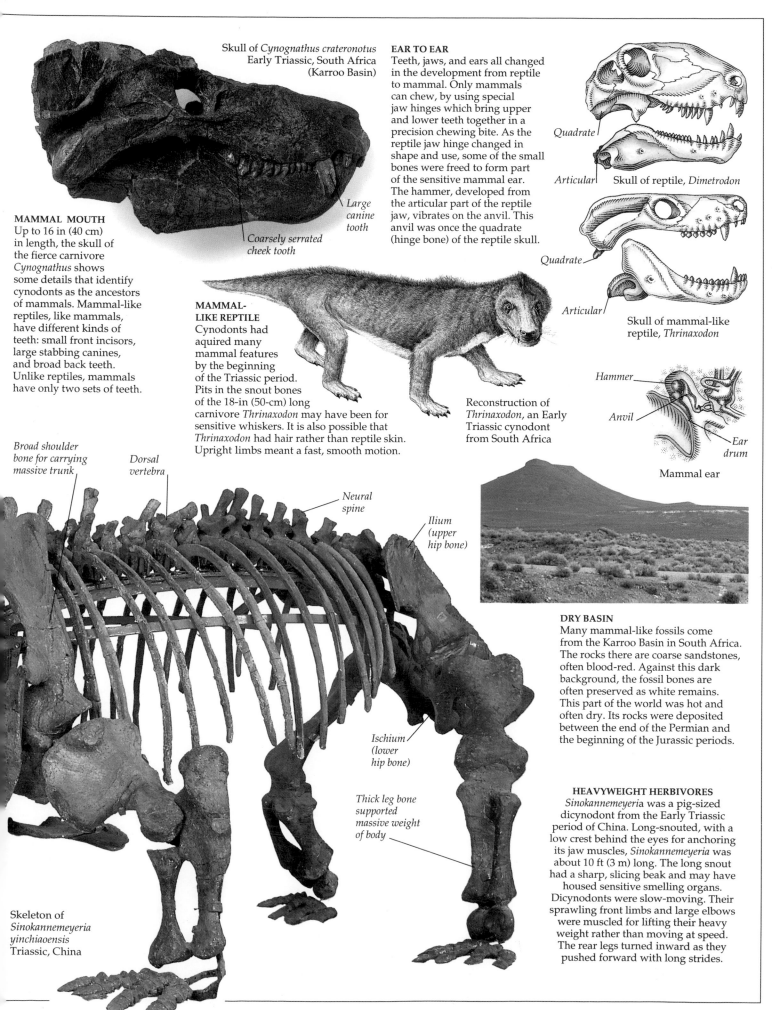

Skull of *Cynognathus crateronotus*
Early Triassic, South Africa
(Karroo Basin)

MAMMAL MOUTH
Up to 16 in (40 cm) in length, the skull of the fierce carnivore *Cynognathus* shows some details that identify cynodonts as the ancestors of mammals. Mammal-like reptiles, like mammals, have different kinds of teeth: small front incisors, large stabbing canines, and broad back teeth. Unlike reptiles, mammals have only two sets of teeth.

Large canine tooth

Coarsely serrated cheek tooth

EAR TO EAR
Teeth, jaws, and ears all changed in the development from reptile to mammal. Only mammals can chew, by using special jaw hinges which bring upper and lower teeth together in a precision chewing bite. As the reptile jaw hinge changed in shape and use, some of the small bones were freed to form part of the sensitive mammal ear. The hammer, developed from the articular part of the reptile jaw, vibrates on the anvil. This anvil was once the quadrate (hinge bone) of the reptile skull.

Quadrate

Articular Skull of reptile, *Dimetrodon*

Quadrate

Articular Skull of mammal-like reptile, *Thrinaxodon*

Hammer

Anvil

Ear drum

Mammal ear

MAMMAL-LIKE REPTILE
Cynodonts had aquired many mammal features by the beginning of the Triassic period. Pits in the snout bones of the 18-in (50-cm) long carnivore *Thrinaxodon* may have been for sensitive whiskers. It is also possible that *Thrinaxodon* had hair rather than reptile skin. Upright limbs meant a fast, smooth motion.

Reconstruction of *Thrinaxodon*, an Early Triassic cynodont from South Africa

Broad shoulder bone for carrying massive trunk

Dorsal vertebra

Neural spine

Ilium (upper hip bone)

DRY BASIN
Many mammal-like fossils come from the Karroo Basin in South Africa. The rocks there are coarse sandstones, often blood-red. Against this dark background, the fossil bones are often preserved as white remains. This part of the world was hot and often dry. Its rocks were deposited between the end of the Permian and the beginning of the Jurassic periods.

Ischium (lower hip bone)

Thick leg bone supported massive weight of body

HEAVYWEIGHT HERBIVORES
Sinokannemeyeria was a pig-sized dicynodont from the Early Triassic period of China. Long-snouted, with a low crest behind the eyes for anchoring its jaw muscles, *Sinokannemeyeria* was about 10 ft (3 m) long. The long snout had a sharp, slicing beak and may have housed sensitive smelling organs. Dicynodonts were slow-moving. Their sprawling front limbs and large elbows were muscled for lifting their heavy weight rather than moving at speed. The rear legs turned inward as they pushed forward with long strides.

Skeleton of *Sinokannemeyeria yinchiaoensis*
Triassic, China

The age of the dinosaurs

GIANT PREDATORS, lumbering plant processors, agile browsers, and pack hunters, dinosaurs occupy a grand place in our knowledge of prehistoric life, featuring in movies, on television, in books, on cereal boxes, and in toy stores. These amazing reptiles are split into two groups: saurischian (reptile-hipped) and ornithischian (bird-hipped). Their 160-million-year reign left fossilized clues to how they lived, but much remains to learn about these specialized terrestrial reptiles. Dinosaurs appeared about 228 million years ago, and quickly dominated life on land. Their ability to stand upright and move efficiently helped them become versatile and adaptable but did not save them from extinction 65 million years ago.

Skull of *Eoraptor* (228 mya), Argentina

FIRST SIGNS
Eoraptor is now recognized as the oldest known dinosaur. Discovered in Argentina in 1992, this small 228-million-year-old dinosaur seems to have been a carnivore (meat eater), perhaps the first in a long line of dinosaur terrors.

PROBLEM POSER
Herrerasaurus, also from Argentina, has been described as a theropod dinosaur (a saurischian two-legged carnivore). Some scientists believe that unusual features of its pelvis and feet mean that it cannot be called either saurischian or ornithischian.

Model of 10-ft (3-m) long *Herrerasaurus* (228 mya) Argentina

DINOSAUR RECORDS
Dinosaurs hold the record as the largest of all known animals. *Diplodocus* was one of the longest at 89 ft (27 m) long. It was one of the sauropod dinosaurs, huge four-legged herbivores (plant eaters) with long necks and tails. Sauropods were saurischian dinosaurs, with the two lower bones of the pelvis (the pubis and the ischium) pointing in opposite directions below the pelvic upper bone (the ilium).

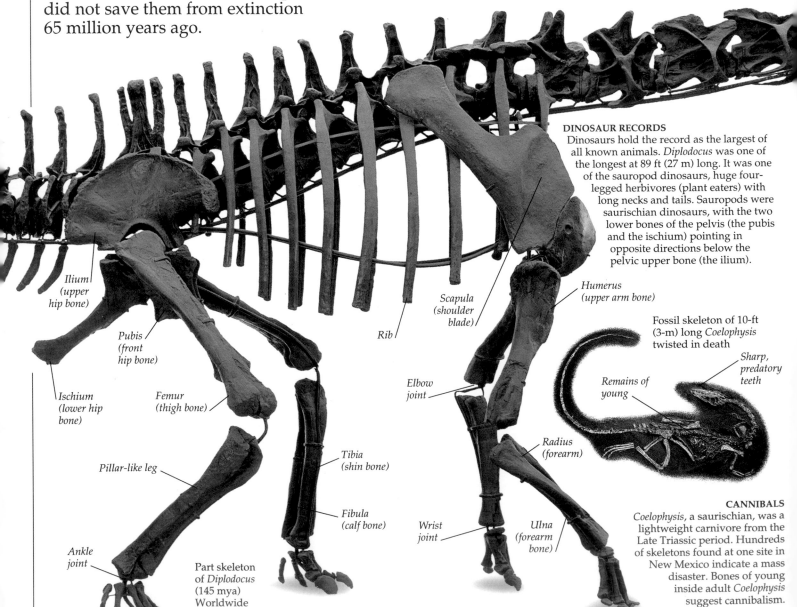

Ilium (upper hip bone)

Scapula (shoulder blade)

Rib

Humerus (upper arm bone)

Pubis (front hip bone)

Elbow joint

Fossil skeleton of 10-ft (3-m) long *Coelophysis* twisted in death

Sharp, predatory teeth

Remains of young

Ischium (lower hip bone)

Femur (thigh bone)

Pillar-like leg

Tibia (shin bone)

Radius (forearm)

Fibula (calf bone)

Wrist joint

Ulna (forearm bone)

Ankle joint

Part skeleton of *Diplodocus* (145 mya) Worldwide

CANNIBALS
Coelophysis, a saurischian, was a lightweight carnivore from the Late Triassic period. Hundreds of skeletons found at one site in New Mexico indicate a mass disaster. Bones of young inside adult *Coelophysis* suggest cannibalism.

Model of 13-ft (4-m) long
Ornithosuchus, Late Triassic
Scotland

Whiplash tail tip

Model of
Diplodocus

Long tail

Body raised
on erect legs

JURASSIC FEATURE
Diplodocus lived
about 145 mya, near
the end of the Jurassic period. Giant plant-eating
sauropod dinosaurs were typical of the Jurassic.
Their fossils are found around the world, in North
and South America, Australia, Europe, and China.

LEG POWER
Ornithosuchus, from the Late
Triassic period of Scotland,
belonged to the archosaurs,
which include dinosaurs as
well as crocodiles and birds.
Ornithosuchus itself was not a
dinosaur but a member of a
primitive archosaur group,
that led to dinosaurs. It was
a carnivore and walked on four
legs. When running, it rose up
on its hind legs, which were
similar to the legs of dinosaurs.

Spine
for muscle
attachment

Hollowed-out
area made
neck bones
light

Powerful
joint between
neck bones

Small head
in relation
to body

Mandible
(lower jaw)

Neck
bone

Neck rib
for muscle
attachment

Heavy,
stiff tail

Beak

Hip
socket

Tail
vertebra

Ilium

Back teeth
chewed
plants

Ischium

Pubis

Model of 30-ft (9-m)
long *Iguanodon* on all
fours, (140–110 mya)
Europe and U.S.

Thumb
spike

Hooflike
nail

Eye
socket

ORNITHISCHIAN PELVIS
Ornithischian dinosaurs had their pelvic
bones arranged with both the pubis and
ischium pointing down and back. They
sometimes developed an extension of the
pubis in front. They were all herbivores
with beaked mouths. *Iguanodon* is among
the best known and best researched by
paleontologists. It could walk on two or
four legs, thereby spreading its weight.

Toothless
lower jaw

ODD-TOOTHED HERBIVORE
Heterodontosaurus was one of the
smallest ornithischian dinosaurs and
one of the first, living about 205 mya.
Heterodontosaurus can be distinguished
from all other dinosaurs by its high-
crowned, plant-grinding back teeth
and its long, spiked canine teeth.

Fossil skeleton
of 4-ft (1.2-m) long
Heterodontosaurus
(205 mya), South Africa

Back
vertebra

Neck
vertebra

Skeleton of
19-ft 7 in (6-m)
long *Gallimimus*
(73 mya), Mongolia

Ilium

Hip
socket

Tail vertebra

Ischium

OSTRICH DINOSAURS
Described as bird mimics because of their
ostrich-like shape, the ornithomimosaurs
were fast-running saurischians. Beaked and
toothless, *Gallimimus* was one of the fastest
runners of the dinosaur world, reaching
speeds of 30 mph (50 km/h). Like birds,
Gallimimus ran on its toes, its long foot-bones
increasing the length of the legs. Its tail was
used as a steadying rudder in high-speed runs.
The vertebrae in the back half of the tail were
stiffly locked together, and a shift to left or
right could help keep balance and change direction. The forelimbs were not used as legs but
as slim and far-reaching arms. Long clawed
fingers could grasp insect and animal prey.

Pubis

Humerus

Femur

Knee joint

Radius

Tibia

Ulna

Fibula

Wrist
joint

Ankle
joint

Long finger for
grasping prey

DESERT DISCOVERIES
Nearly all ornithomimosaurs have been
found in western North America and
eastern Asia. They date from the Cretaceous period. *Gallimimus* was discovered
in the late 1960s on an expedition to the
Gobi Desert, where many spectacular
dinosaur remains have been found.

Saurischians

DINOSAURS ARE EXTINCT. Their bones, teeth, eggshells, footprints, and skin impressions are all that they left behind. Taxonomists use this evidence to work out which dinosaurs belong in the same groups. Details such as the number of toe bones or the arrangement of skull bones may differ from one group to another. Special features shared by several groups show which dinosaurs might be related in families. Saurischians (reptile-hipped dinosaurs) are divided into two main groups, the meat-eating theropods (such as *Allosaurus*) and the plant-eating sauropods (such as *Camarasaurus*).

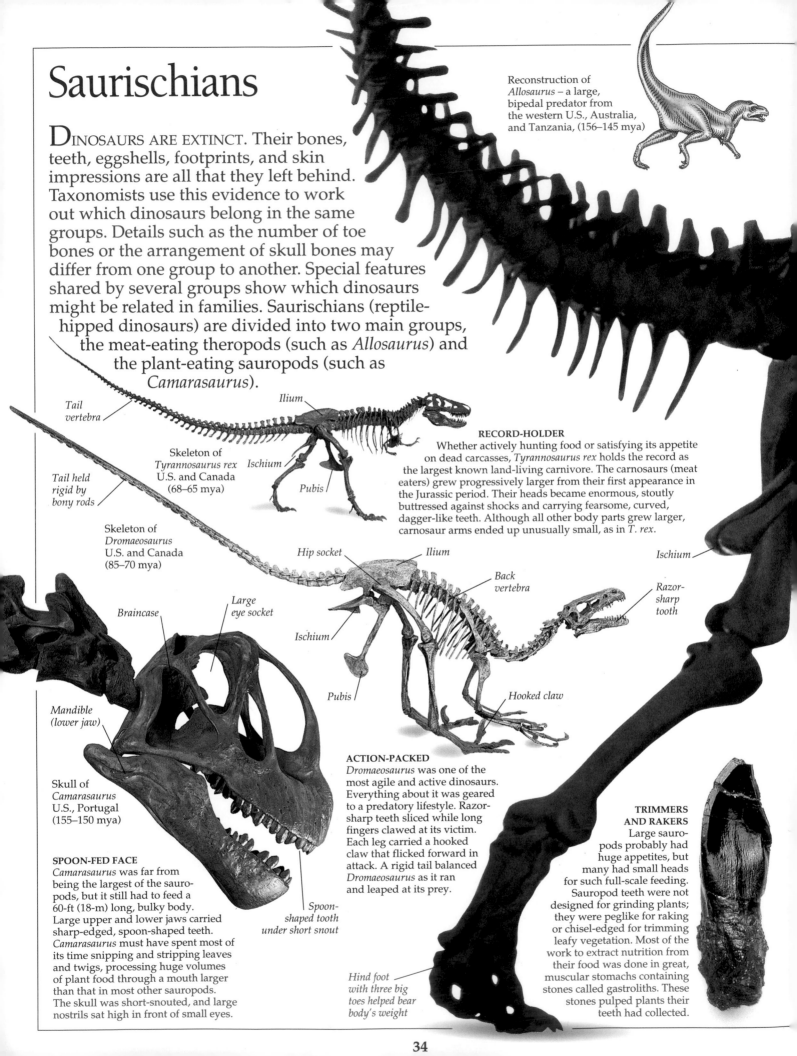

Reconstruction of *Allosaurus* – a large, bipedal predator from the western U.S., Australia, and Tanzania, (156–145 mya)

Tail vertebra

Skeleton of *Tyrannosaurus rex* U.S. and Canada (68–65 mya)

Ilium

Ischium

Pubis

Tail held rigid by bony rods

Skeleton of *Dromaeosaurus* U.S. and Canada (85–70 mya)

Hip socket

Ilium

Back vertebra

Ischium

Ischium

Pubis

Hooked claw

Razor-sharp tooth

Braincase

Large eye socket

Mandible (lower jaw)

Skull of *Camarasaurus* U.S., Portugal (155–150 mya)

Spoon-shaped tooth under short snout

Hind foot with three big toes helped bear body's weight

RECORD-HOLDER
Whether actively hunting food or satisfying its appetite on dead carcasses, *Tyrannosaurus rex* holds the record as the largest known land-living carnivore. The carnosaurs (meat eaters) grew progressively larger from their first appearance in the Jurassic period. Their heads became enormous, stoutly buttressed against shocks and carrying fearsome, curved, dagger-like teeth. Although all other body parts grew larger, carnosaur arms ended up unusually small, as in *T. rex*.

SPOON-FED FACE
Camarasaurus was far from being the largest of the sauropods, but it still had to feed a 60-ft (18-m) long, bulky body. Large upper and lower jaws carried sharp-edged, spoon-shaped teeth. *Camarasaurus* must have spent most of its time snipping and stripping leaves and twigs, processing huge volumes of plant food through a mouth larger than that in most other sauropods. The skull was short-snouted, and large nostrils sat high in front of small eyes.

ACTION-PACKED
Dromaeosaurus was one of the most agile and active dinosaurs. Everything about it was geared to a predatory lifestyle. Razor-sharp teeth sliced while long fingers clawed at its victim. Each leg carried a hooked claw that flicked forward in attack. A rigid tail balanced *Dromaeosaurus* as it ran and leaped at its prey.

TRIMMERS AND RAKERS
Large sauropods probably had huge appetites, but many had small heads for such full-scale feeding. Sauropod teeth were not designed for grinding plants; they were peglike for raking or chisel-edged for trimming leafy vegetation. Most of the work to extract nutrition from their food was done in great, muscular stomachs containing stones called gastroliths. These stones pulped plants their teeth had collected.

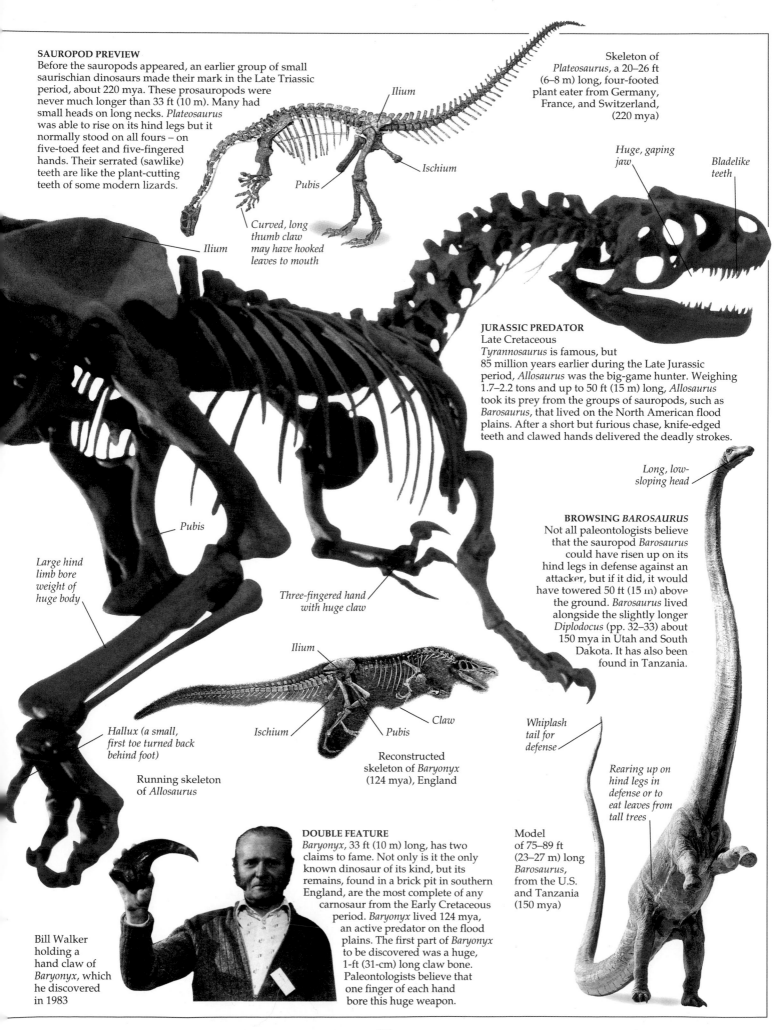

SAUROPOD PREVIEW
Before the sauropods appeared, an earlier group of small saurischian dinosaurs made their mark in the Late Triassic period, about 220 mya. These prosauropods were never much longer than 33 ft (10 m). Many had small heads on long necks. *Plateosaurus* was able to rise on its hind legs but it normally stood on all fours – on five-toed feet and five-fingered hands. Their serrated (sawlike) teeth are like the plant-cutting teeth of some modern lizards.

Skeleton of *Plateosaurus*, a 20–26 ft (6–8 m) long, four-footed plant eater from Germany, France, and Switzerland, (220 mya)

Ilium

Ischium

Pubis

Huge, gaping jaw

Bladelike teeth

Curved, long thumb claw may have hooked leaves to mouth

Ilium

JURASSIC PREDATOR
Late Cretaceous *Tyrannosaurus* is famous, but 85 million years earlier during the Late Jurassic period, *Allosaurus* was the big-game hunter. Weighing 1.7–2.2 tons and up to 50 ft (15 m) long, *Allosaurus* took its prey from the groups of sauropods, such as *Barosaurus*, that lived on the North American flood plains. After a short but furious chase, knife-edged teeth and clawed hands delivered the deadly strokes.

Pubis

Large hind limb bore weight of huge body

Three-fingered hand with huge claw

Long, low-sloping head

BROWSING *BAROSAURUS*
Not all paleontologists believe that the sauropod *Barosaurus* could have risen up on its hind legs in defense against an attacker, but if it did, it would have towered 50 ft (15 m) above the ground. *Barosaurus* lived alongside the slightly longer *Diplodocus* (pp. 32–33) about 150 mya in Utah and South Dakota. It has also been found in Tanzania.

Ilium

Hallux (a small, first toe turned back behind foot)

Ischium

Pubis

Claw

Reconstructed skeleton of *Baryonyx* (124 mya), England

Whiplash tail for defense

Rearing up on hind legs in defense or to eat leaves from tall trees

Running skeleton of *Allosaurus*

DOUBLE FEATURE
Baryonyx, 33 ft (10 m) long, has two claims to fame. Not only is it the only known dinosaur of its kind, but its remains, found in a brick pit in southern England, are the most complete of any carnosaur from the Early Cretaceous period. *Baryonyx* lived 124 mya, an active predator on the flood plains. The first part of *Baryonyx* to be discovered was a huge, 1-ft (31-cm) long claw bone. Paleontologists believe that one finger of each hand bore this huge weapon.

Model of 75–89 ft (23–27 m) long *Barosaurus*, from the U.S. and Tanzania (150 mya)

Bill Walker holding a hand claw of *Baryonyx*, which he discovered in 1983

Ornithischians

ORNITHISCHIAN (BIRD-HIPPED) dinosaurs had many features in common. Large or small, they were all herbivores, feeding on leaves, fruits, seeds, even conifer needles. Teeth were arranged for slicing and grinding, but ornithischians were toothless at the front of their mouths. A sharp-edged beak did the cutting and tearing instead. All ornithischians had an extra bone, the predentary, at the front of the lower jaw. Another bone unique to ornithischians was the palpebral, in the eyelid. Many ornithischians, like the hadrosaurs, were somewhat bipedal; they walked on their hind legs.

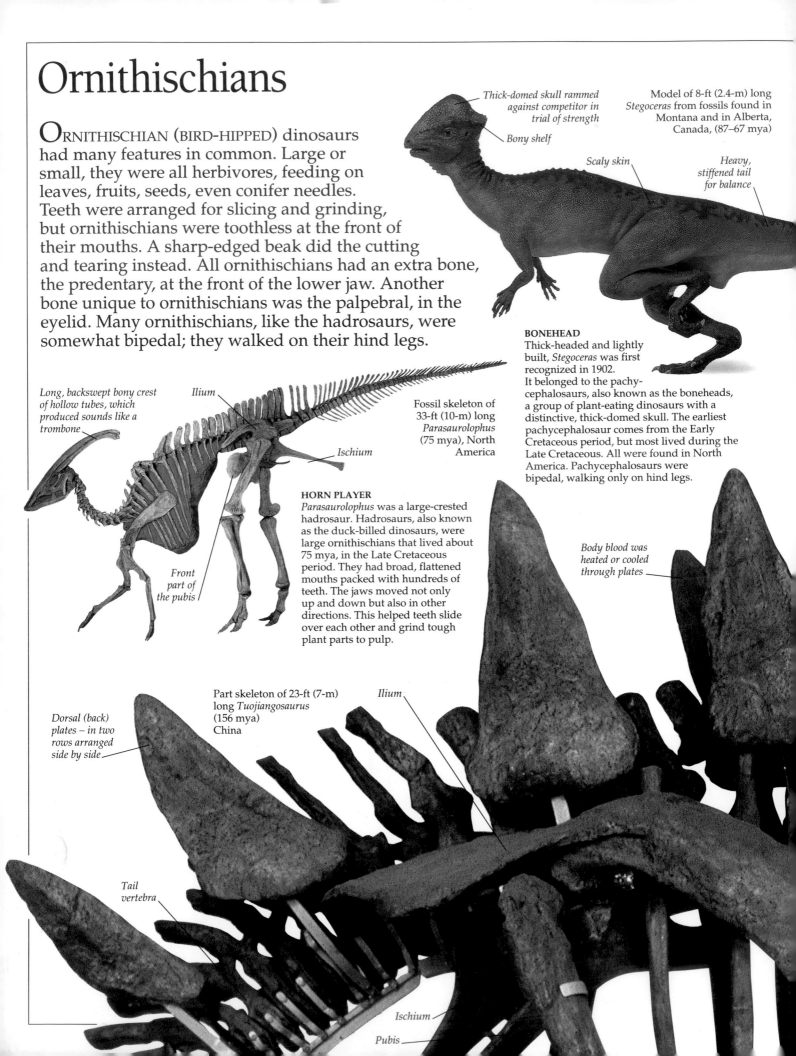

Thick-domed skull rammed against competitor in trial of strength

Bony shelf

Model of 8-ft (2.4-m) long *Stegoceras* from fossils found in Montana and in Alberta, Canada, (87–67 mya)

Scaly skin

Heavy, stiffened tail for balance

BONEHEAD
Thick-headed and lightly built, *Stegoceras* was first recognized in 1902. It belonged to the pachycephalosaurs, also known as the boneheads, a group of plant-eating dinosaurs with a distinctive, thick-domed skull. The earliest pachycephalosaur comes from the Early Cretaceous period, but most lived during the Late Cretaceous. All were found in North America. Pachycephalosaurs were bipedal, walking only on hind legs.

Long, backswept bony crest of hollow tubes, which produced sounds like a trombone

Ilium

Fossil skeleton of 33-ft (10-m) long *Parasaurolophus* (75 mya), North America

Ischium

Front part of the pubis

HORN PLAYER
Parasaurolophus was a large-crested hadrosaur. Hadrosaurs, also known as the duck-billed dinosaurs, were large ornithischians that lived about 75 mya, in the Late Cretaceous period. They had broad, flattened mouths packed with hundreds of teeth. The jaws moved not only up and down but also in other directions. This helped teeth slide over each other and grind tough plant parts to pulp.

Body blood was heated or cooled through plates

Dorsal (back) plates – in two rows arranged side by side

Part skeleton of 23-ft (7-m) long *Tuojiangosaurus* (156 mya) China

Ilium

Tail vertebra

Ischium

Pubis

Horny beak

Extended foot-bones for running

Fast-moving 7.5-ft (2.3-m) *Hypsilophodon*, (120 mya) Europe and U.S.

QUICK MOVER
One of the smallest ornithischians, *Hypsilophodon* was a delicately built herbivore. Hypsilophodontids have been found in Australia, Europe, North America, and China, spreading across continents when they were land-linked. *Hypsilophodon* used its nimble legs to quickly race away from danger.

BIG HEADS
Ceratopians, such as *Triceratops*, were the big heads of the dinosaur world. Known as the horned dinosaurs, their brow horns, nose horns, frills of bone, and rims of spikes made an impressive head display that warned off predators. Ceratopians had an extra bone, the rostral bone, at the tip of a beaked snout. The beak snipped and tore at plants while the teeth of the powerful jaws sheared them. The horned dinosaurs lived at the end of the Cretaceous period.

Frill

Long brow-horn

Short nose-horn

Rostral bone

Predentary bone

Skull and lower jaw of *Triceratops* (67–65 mya) North America

ARMORED ANKYLOSAUR
Quadrupedal *Euoplocephalus*, a plant-eating ornithischian, was heavy and lumbering on its four feet. An ankylosaur, it relied on bony armor to fend off predators. It had thick bony plates and short spikes over its back, and even specially thickened bone on its head. Ankylosaurs first appeared about 188 mya. Some were among the last dinosaurs alive.

Shoulder spike gave extra protection

Claw

Swinging tail-club, made of interlocking vertebrae

Well-protected 23-ft (7-m) long *Euoplocephalus*, (70 mya) North America

Plates acted like solar panels and radiators for body

SPIKY SIGNPOSTS
Standing upright like signposts, the bony back-plates of stegosaurs were an easily identified feature. Two rows of plates ran from neck to tail. The tails of some stegosaurs ended in sharp, bony spines. The last known stegosaur came from India's Late Cretaceous period about 88 mya, but most fossil remains are from the Late Jurassic, 157–145 mya. Quadrupedal plant-eater *Tuojiangosaurus*, from China, had tall, narrow, spike-like back-plates.

AN AMERICAN STEGOSAUR
The North American *Stegosaurus* (found in Wyoming, Colorado, and Utah) is probably the most famous of the stegosaurs. At 30 ft (9 m) long, it was the largest, and its two rows of broad back-plates were staggered rather than arranged side by side. A herbivore of 150 mya, *Stegosaurus*'s diet consisted of low-growing plants, which it chewed with small, serrated, leaflike teeth.

Shoulder blade

Reptiles at sea

AFTER LIVING ON LAND for over 80 million years, various reptiles independently adopted a fully aquatic life and dominated the Mesozoic seas as swimming predators. As air-breathing reptiles, they had to surface to fill their lungs, but many features of sight, smell, and respiration became adapted to a marine environment. In marine reptiles, the bones of walking limbs and feet developed into a variety of paddle shapes used as underwater oars or wings. Some reptiles, like turtles, still hauled themselves onto land to lay eggs; underwater, the embryos would suffocate for lack of oxygen. One group of marine reptiles, the ichthyosaurs, cut themselves off from land completely and gave birth to live young (pp. 52–53) in the water.

COMMERCIAL COLLECTOR
Mary Anning (1799–1847) was one of the first commercial fossil collectors. She lived on England's south coast in the seaside town of Lyme Regis, where the rocks are Jurassic in age, and collected fossils to be sold in her father's curiosity shop. She is credited with finding her first ichthyosaur specimen at the age of 11. Many museums have specimens of marine reptiles that Mary Anning discovered.

Skeleton of *Cryptoclidus eurymerus* Mid Jurassic England

Flat-ended femur joined with pelvis

Flat, rigid pelvic girdle

Belly ribs strengthened underside of body

Huge flipper made up of five elongated toes

Most flexible vertebrae in neck

MARINE FLYER
Plesiosaurs came in two sizes: short-necked, large-headed pliosaurids and long-necked, small-headed plesiosaurids. The jaws of long-necked *Cryptoclidus* had long, spiked teeth that interlocked like two combs when they closed around a fish. The shoulder and pelvic bones were massive, forming a rigid anchorage for the tapered flippers. Some scientists believe that the flippers paddled to and fro, while others think they were used for underwater flight, as with penguins and some turtles.

ACTION-PACKED
Although this scene is rather fanciful, the Mesozoic era was a time when reptiles ruled the sea, land, and air. The largest marine reptiles preyed on other reptiles, while smaller swimmers fed on fish or specialized in squid or mollusks. Turtles and crocodiles survived the extinction that befell most marine reptiles.

Stenopterygius quadricissus Early Jurassic, Germany

Back paddle of *Protostega gigas*, an ancient turtle

STREAMLINED SWIMMER
This skin impression of *Stenopterygius* displays the features of the highly effective ichthyosaurs, so perfectly shaped for life at sea. Streamlined for speed, they were propelled by the sideways motion of a vertical, sharklike tail. Their flippers were used to stabilize and steer. A long, tooth-lined snout suggests comparisons with modern dolphins, not just in their diet of squid and fish, but also in their active leaping above the waves. A large ring of bones surrounded huge eyes, suggesting that *Stenopterygius* had keen vision in the deep, dark waters.

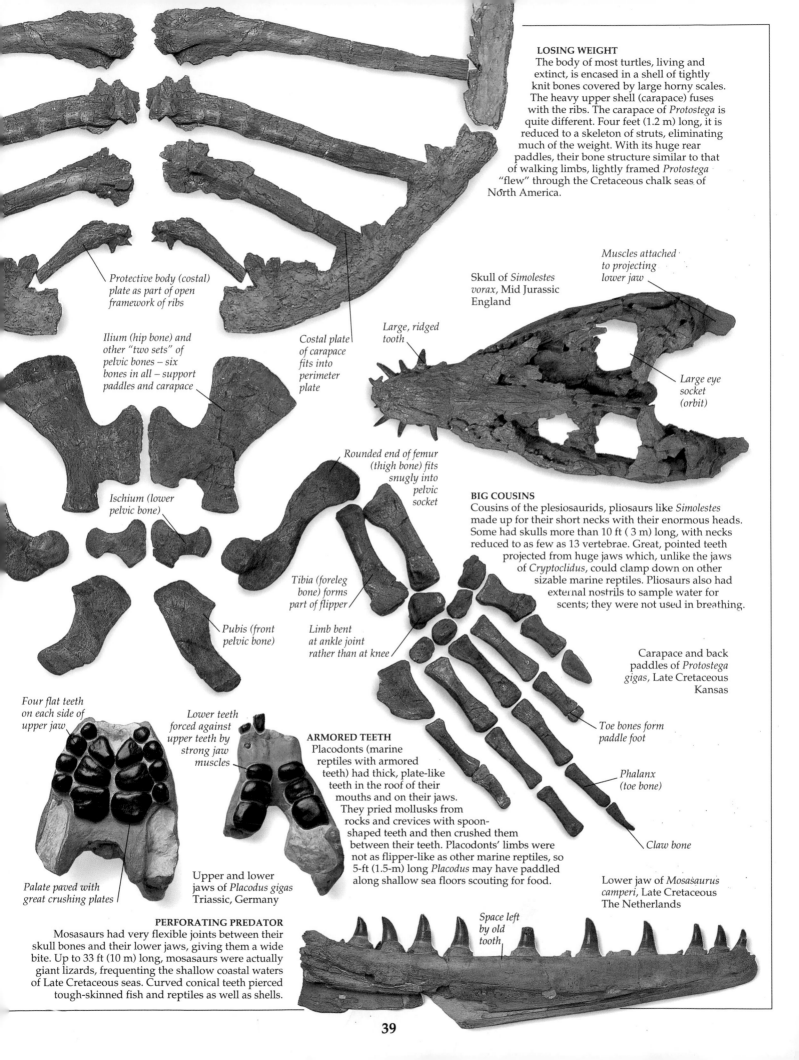

LOSING WEIGHT
The body of most turtles, living and extinct, is encased in a shell of tightly knit bones covered by large horny scales. The heavy upper shell (carapace) fuses with the ribs. The carapace of *Protostega* is quite different. Four feet (1.2 m) long, it is reduced to a skeleton of struts, eliminating much of the weight. With its huge rear paddles, their bone structure similar to that of walking limbs, lightly framed *Protostega* "flew" through the Cretaceous chalk seas of North America.

Protective body (costal) plate as part of open framework of ribs

Ilium (hip bone) and other "two sets" of pelvic bones – six bones in all – support paddles and carapace

Costal plate of carapace fits into perimeter plate

Ischium (lower pelvic bone)

Rounded end of femur (thigh bone) fits snugly into pelvic socket

Tibia (foreleg bone) forms part of flipper

Pubis (front pelvic bone)

Limb bent at ankle joint rather than at knee

Muscles attached to projecting lower jaw

Skull of Simolestes vorax, Mid Jurassic England

Large, ridged tooth

Large eye socket (orbit)

BIG COUSINS
Cousins of the plesiosaurids, pliosaurs like *Simolestes* made up for their short necks with their enormous heads. Some had skulls more than 10 ft (3 m) long, with necks reduced to as few as 13 vertebrae. Great, pointed teeth projected from huge jaws which, unlike the jaws of *Cryptoclidus*, could clamp down on other sizable marine reptiles. Pliosaurs also had external nostrils to sample water for scents; they were not used in breathing.

Carapace and back paddles of Protostega gigas, Late Cretaceous Kansas

Toe bones form paddle foot

Phalanx (toe bone)

Claw bone

ARMORED TEETH
Placodonts (marine reptiles with armored teeth) had thick, plate-like teeth in the roof of their mouths and on their jaws. They pried mollusks from rocks and crevices with spoon-shaped teeth and then crushed them between their teeth. Placodonts' limbs were not as flipper-like as other marine reptiles, so 5-ft (1.5-m) long *Placodus* may have paddled along shallow sea floors scouting for food.

Four flat teeth on each side of upper jaw

Lower teeth forced against upper teeth by strong jaw muscles

Palate paved with great crushing plates

Upper and lower jaws of *Placodus gigas* Triassic, Germany

Lower jaw of *Mosasaurus camperi*, Late Cretaceous The Netherlands

PERFORATING PREDATOR
Mosasaurs had very flexible joints between their skull bones and their lower jaws, giving them a wide bite. Up to 33 ft (10 m) long, mosasaurs were actually giant lizards, frequenting the shallow coastal waters of Late Cretaceous seas. Curved conical teeth pierced tough-skinned fish and reptiles as well as shells.

Space left by old tooth

Flying reptiles

THE LARGEST FLIERS EVER were pterosaurs, which took to the skies in the Triassic period. These archosaur reptiles, closely related to dinosaurs, had wings of reinforced skin stretched across one long finger and powerful muscles. Although pterosaurs had lightweight frames of hollow, air-filled bone, some weighed up to 220 lb (100 kg). There are two kinds of pterosaurs. The earlier were long-tailed, short-headed rhamphorhynchoids (like *Rhamphorhynchus*). These died out at the end of the Jurassic period, but not before the pterodactyloids had appeared. These short-tailed, long-headed pterosaurs survived until extinction at the end of the Cretaceous period.

SEA FOOD
Fossilized food has been found in the stomachs of several pterosaur remains. Most pterosaur fossils (such as Early Cretaceous *Anhanguera* from Brazil) are found in rocks deposited in shallow seas and some contain fish fossils. *Anhanguera*'s long jaws were ideal for scooping up a slippery catch as it flew low over the water. Crests on its beak helped stabilize its head as it dipped into the sea.

FLIGHT ENGINEER
How *Quetzalcoatlus*, with a wingspan of 40 ft (12 m) and weighing as much as 190 lb (86 kg), could carry itself through the air is a miracle of natural engineering. Named after a feathered Aztec god, the long-necked and toothless *Quetzalcoatlus* is the largest pterosaur known. It was far larger than today's giant albatross.

Typical long neck of a pterodactyloid

Model of *Quetzalcoatlus* (65 mya). Its fossil remains were first found in 1971 in Texas.

An enormous 40-ft (12-m) wingspan supported Quetzalcoatlus *in the air*

Wings of Criorhynchus would have been folded back when not flying

Small fragments of bones are pieced together to build complete finger

Only the first joint of the wing finger could bend; the rest of the finger remained rigid

Hollow finger bone reduced weight of wing

Fourth finger

Wing bone of *Rhamphorhynchus* (160–144 mya), Germany

Flight muscles attached to a large crest on the humerus

WING FIBERS
As if stretched out to show its papery fragility, the wrinkled wing of *Rhamphorhynchus* reveals its secret strength. The long fourth finger leads the wing edge. Fine, tough fibers reinforce the membrane of skin. These fibers helped stretch and strengthen the skin to withstand the strain as the wing flapped.

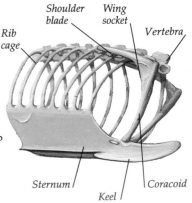

Shoulder blade *Wing socket*

Rib cage

Vertebra

Sternum *Keel* *Coracoid*

Breastbone of *Rhamphorhynchus*

Wing membrane

Terminal vane (rudder) on tail

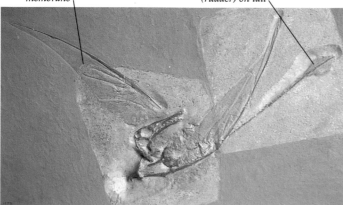

FALLEN FLIER
Famous for its many superb fossils, the Solnhofen limestone (pp. 42–43) seems to have captured this *Rhamphorhynchus* as it fell to its death (left). A common pterosaur from the Jurassic period, with a wingspan of up to 5 ft 9 in (1.75 m), *Rhamphorhynchus* stabilized and perhaps steered itself with its long-tailed rudder. Flight muscles were attached to a large-keeled breastbone (sternum) as in flying birds. Ribs joined the breastbone to make a strong frame. The shoulder bones (shoulder blade and coracoid) braced the powerful wing movements against the body frame.

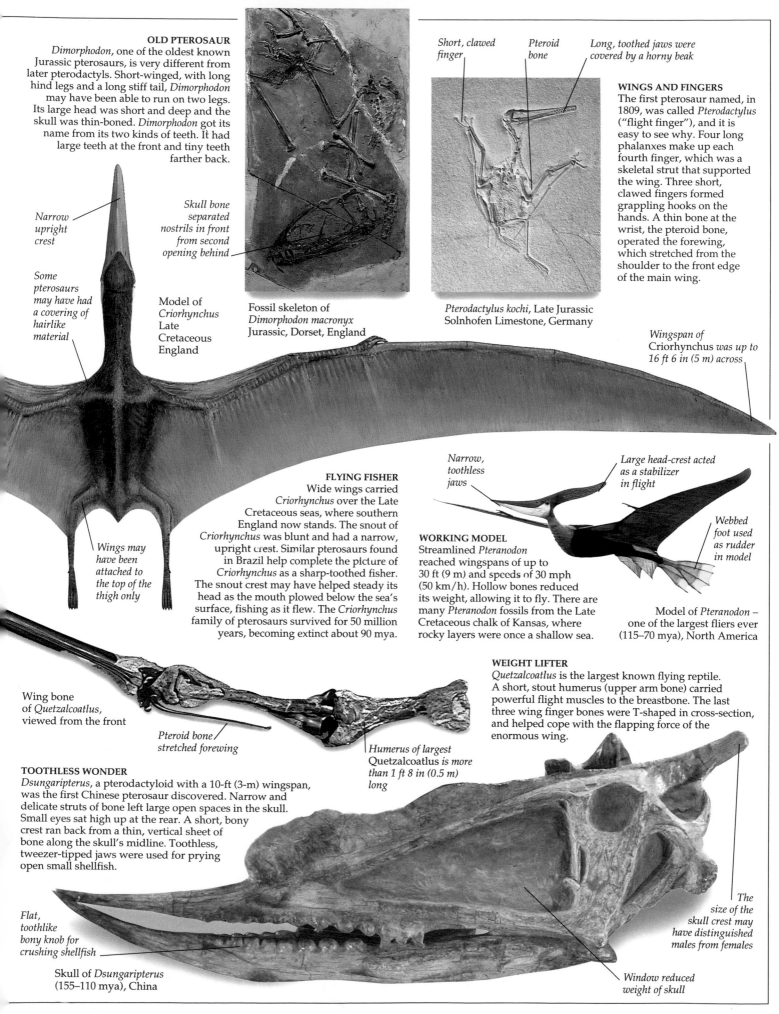

OLD PTEROSAUR
Dimorphodon, one of the oldest known Jurassic pterosaurs, is very different from later pterodactyls. Short-winged, with long hind legs and a long stiff tail, *Dimorphodon* may have been able to run on two legs. Its large head was short and deep and the skull was thin-boned. *Dimorphodon* got its name from its two kinds of teeth. It had large teeth at the front and tiny teeth farther back.

Narrow upright crest

Some pterosaurs may have had a covering of hairlike material

Model of *Criorhynchus* Late Cretaceous England

Skull bone separated nostrils in front from second opening behind

Fossil skeleton of *Dimorphodon macronyx* Jurassic, Dorset, England

Short, clawed finger

Pteroid bone

Long, toothed jaws were covered by a horny beak

WINGS AND FINGERS
The first pterosaur named, in 1809, was called *Pterodactylus* ("flight finger"), and it is easy to see why. Four long phalanxes make up each fourth finger, which was a skeletal strut that supported the wing. Three short, clawed fingers formed grappling hooks on the hands. A thin bone at the wrist, the pteroid bone, operated the forewing, which stretched from the shoulder to the front edge of the main wing.

Pterodactylus kochi, Late Jurassic Solnhofen Limestone, Germany

Wingspan of Criorhynchus was up to 16 ft 6 in (5 m) across

Narrow, toothless jaws

Large head-crest acted as a stabilizer in flight

FLYING FISHER
Wide wings carried *Criorhynchus* over the Late Cretaceous seas, where southern England now stands. The snout of *Criorhynchus* was blunt and had a narrow, upright crest. Similar pterosaurs found in Brazil help complete the picture of *Criorhynchus* as a sharp-toothed fisher. The snout crest may have helped steady its head as the mouth plowed below the sea's surface, fishing as it flew. The *Criorhynchus* family of pterosaurs survived for 50 million years, becoming extinct about 90 mya.

Wings may have been attached to the top of the thigh only

WORKING MODEL
Streamlined *Pteranodon* reached wingspans of up to 30 ft (9 m) and speeds of 30 mph (50 km/h). Hollow bones reduced its weight, allowing it to fly. There are many *Pteranodon* fossils from the Late Cretaceous chalk of Kansas, where rocky layers were once a shallow sea.

Webbed foot used as rudder in model

Model of *Pteranodon* – one of the largest fliers ever (115–70 mya), North America

WEIGHT LIFTER
Quetzalcoatlus is the largest known flying reptile. A short, stout humerus (upper arm bone) carried powerful flight muscles to the breastbone. The last three wing finger bones were T-shaped in cross-section, and helped cope with the flapping force of the enormous wing.

Wing bone of *Quetzalcoatlus*, viewed from the front

Pteroid bone stretched forewing

Humerus of largest Quetzalcoatlus *is more than 1 ft 8 in (0.5 m) long*

TOOTHLESS WONDER
Dsungaripterus, a pterodactyloid with a 10-ft (3-m) wingspan, was the first Chinese pterosaur discovered. Narrow and delicate struts of bone left large open spaces in the skull. Small eyes sat high up at the rear. A short, bony crest ran back from a thin, vertical sheet of bone along the skull's midline. Toothless, tweezer-tipped jaws were used for prying open small shellfish.

Flat, toothlike bony knob for crushing shellfish

Skull of *Dsungaripterus* (155–110 mya), China

The size of the skull crest may have distinguished males from females

Window reduced weight of skull

Early birds

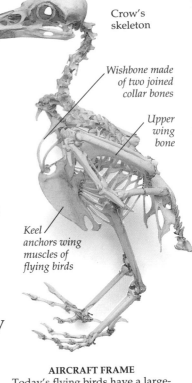

Crow's skeleton

Wishbone made of two joined collar bones

Upper wing bone

Keel anchors wing muscles of flying birds

BIRDS POSSESS ONE FEATURE found in no other animal: feathers. The first fossil evidence of birds was a 147-million-year-old feather. The earliest bird, *Archaeopteryx* ("ancient wing"), has perfectly formed feathers preserved as fossils. But no fossil intermediate between feathers and scales, from which feathers may have developed, has ever been found. Feathers were useful for body insulation before they became useful for flight. Birds are now recognized as the closest living relatives of dinosaurs. Scientists have found 21 features common to both *Archaeopteryx* and small theropod dinosaurs (pp. 32–35). Eight skeletons of *Archaeopteryx* have been found since 1861; they are among the world's most famous and precious fossils.

FLYING FOOD
Birds are not the only animals that fly. The Late Jurassic skies would have had a wide range of insect life, such as this dragonfly (*Libellula* from Solnhofen in Germany), hunting for food in water or on land. Some scientists think *Archaeopteryx* may have used its feathers as fly swatters!

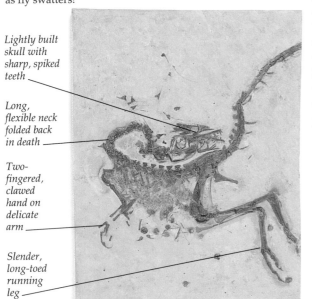

Lightly built skull with sharp, spiked teeth

Long, flexible neck folded back in death

Two-fingered, clawed hand on delicate arm

Slender, long-toed running leg

FEATHERED DINOSAUR
One particular specimen of *Archaeopteryx* that had left few, if any, feather impressions was identified for many years as this dinosaur, *Compsognathus* (found in the same rock deposits). That some dinosaurs and birds are so hard to tell apart reinforces the idea that birds and dinosaurs are closely related.

RARE FOSSILS
Eight specimens of *Archaeopteryx* have been found in quarries of Solnhofen limestone in Germany. The rocks contain fossils of insects, pterosaurs, and dinosaurs also.

BERLIN *ARCHAEOPTERYX*
Discovered in 1877, this specimen of *Archaeopteryx* (now in Berlin's Humboldt Museum in Germany) is the most complete. Perfect feather impressions identify it as a bird. The birdlike foot has three sharply clawed toes in front and a reversed toe behind. Its long, narrow skull has sharp, reptilian teeth but no beak. Feathers fringe a long tailbone, unlike the short, stumpy one of today's birds. The winged arms end in three clawed fingers.

AIRCRAFT FRAME
Today's flying birds have a large-keeled breastbone that anchors strong wing muscles and a wishbone that buttresses the wing joints.

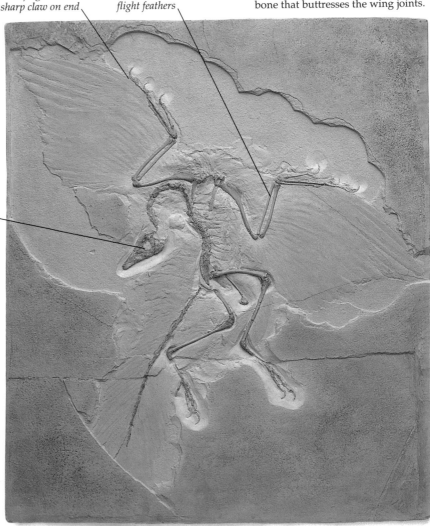

Slim finger with sharp claw on end

Birdlike arm carried flight feathers

Large eye-socket and brain in lightweight skull

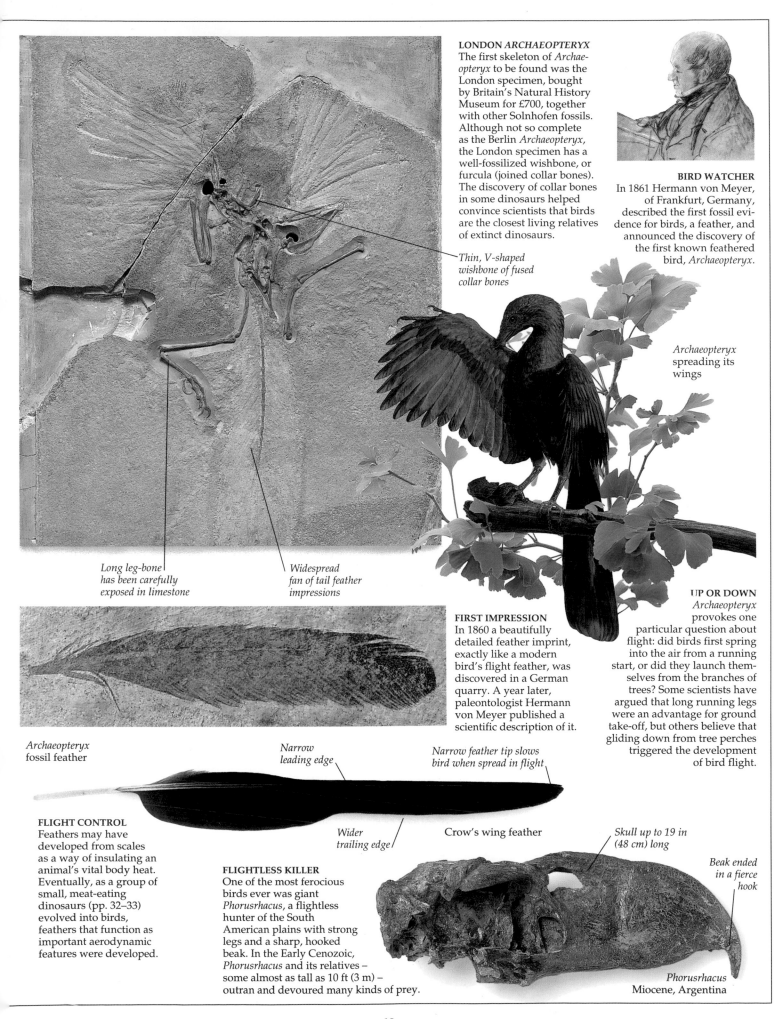

LONDON *ARCHAEOPTERYX*
The first skeleton of *Archaeopteryx* to be found was the London specimen, bought by Britain's Natural History Museum for £700, together with other Solnhofen fossils. Although not so complete as the Berlin *Archaeopteryx*, the London specimen has a well-fossilized wishbone, or furcula (joined collar bones). The discovery of collar bones in some dinosaurs helped convince scientists that birds are the closest living relatives of extinct dinosaurs.

Thin, V-shaped wishbone of fused collar bones

BIRD WATCHER
In 1861 Hermann von Meyer, of Frankfurt, Germany, described the first fossil evidence for birds, a feather, and announced the discovery of the first known feathered bird, *Archaeopteryx*.

Archaeopteryx spreading its wings

Long leg-bone has been carefully exposed in limestone

Widespread fan of tail feather impressions

FIRST IMPRESSION
In 1860 a beautifully detailed feather imprint, exactly like a modern bird's flight feather, was discovered in a German quarry. A year later, paleontologist Hermann von Meyer published a scientific description of it.

UP OR DOWN
Archaeopteryx provokes one particular question about flight: did birds first spring into the air from a running start, or did they launch themselves from the branches of trees? Some scientists have argued that long running legs were an advantage for ground take-off, but others believe that gliding down from tree perches triggered the development of bird flight.

Archaeopteryx fossil feather

Narrow leading edge

Narrow feather tip slows bird when spread in flight

FLIGHT CONTROL
Feathers may have developed from scales as a way of insulating an animal's vital body heat. Eventually, as a group of small, meat-eating dinosaurs (pp. 32–33) evolved into birds, feathers that function as important aerodynamic features were developed.

Wider trailing edge

Crow's wing feather

Skull up to 19 in (48 cm) long

Beak ended in a fierce hook

FLIGHTLESS KILLER
One of the most ferocious birds ever was giant *Phorusrhacus*, a flightless hunter of the South American plains with strong legs and a sharp, hooked beak. In the Early Cenozoic, *Phorusrhacus* and its relatives – some almost as tall as 10 ft (3 m) – outran and devoured many kinds of prey.

Phorusrhacus
Miocene, Argentina

Mammals take over

Warm-blooded mammals have existed for many millions of years. Dwarfed by dinosaurs for 165 million years, mammals survived the mass extinctions at the end of the Cretaceous period. Early mammals may have evaded dinosaur predators partly because they were small (rarely larger than rats) and appeared mainly at night. Dinosaurs had been a varied, successful group of animals, occupying many different habitats. After they became extinct, mammals took over. Two major groups of mammals are alive today, marsupials and placentals, and both give birth to live offspring. Marsupial young are immature at birth, and develop further in the adult's pouch. Placental young are more advanced at birth, having been fed through the placenta inside the mother's womb. Although both groups are found as fossils in Cretaceous rocks, it was the placental mammals that became dominant on Earth.

MILK AND FUR
Fur-coated, and able to convert food to energy quickly, mammals keep their body temperature constant. Newborns grow fast by feeding on nourishing milk from their mother's mammary glands. These specialized sweat glands are unique to mammals.

Foot

Pelvis

Fossil skeleton of *Megazostrodon*

Fur bristles have been fossilized

MINI MAMMAL
Looking like a shrew, *Megazostrodon* from South Africa belonged to a group of tiny mammals, the morganucodontids, of the Late Triassic and Early Jurassic periods. Covered with hair, *Megazostrodon* could maintain an efficient body temperature, unlike the sun-bathing reptiles.

Hair insulates body

Scaly plates on tail

Right hind foot

Long tail helps balance

Model of *Megazostrodon*, one of the first mammals Late Triassic South Africa

An adult *Uintatherium* with a youngster

Pair of large horns at back of skull

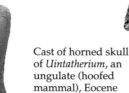

Cast of horned skull of *Uintatherium*, an ungulate (hoofed mammal), Eocene North America

Pair of smaller horns on forehead

Pair of nasal horns

Molar tooth

HORN-HEAD
One of the first large mammals, *Uintatherium* was the size of a rhinoceros. It lived 50 million years ago in North America. *Uintatherium* was a herbivore, using its broad, crested back teeth to slice up stems and leaves. Three pairs of horns adorned its skull, the largest pair at the rear and the smallest on top of the nose. The males also had a pair of large canine teeth; their elaborate skull ornaments may have helped to attract mates.

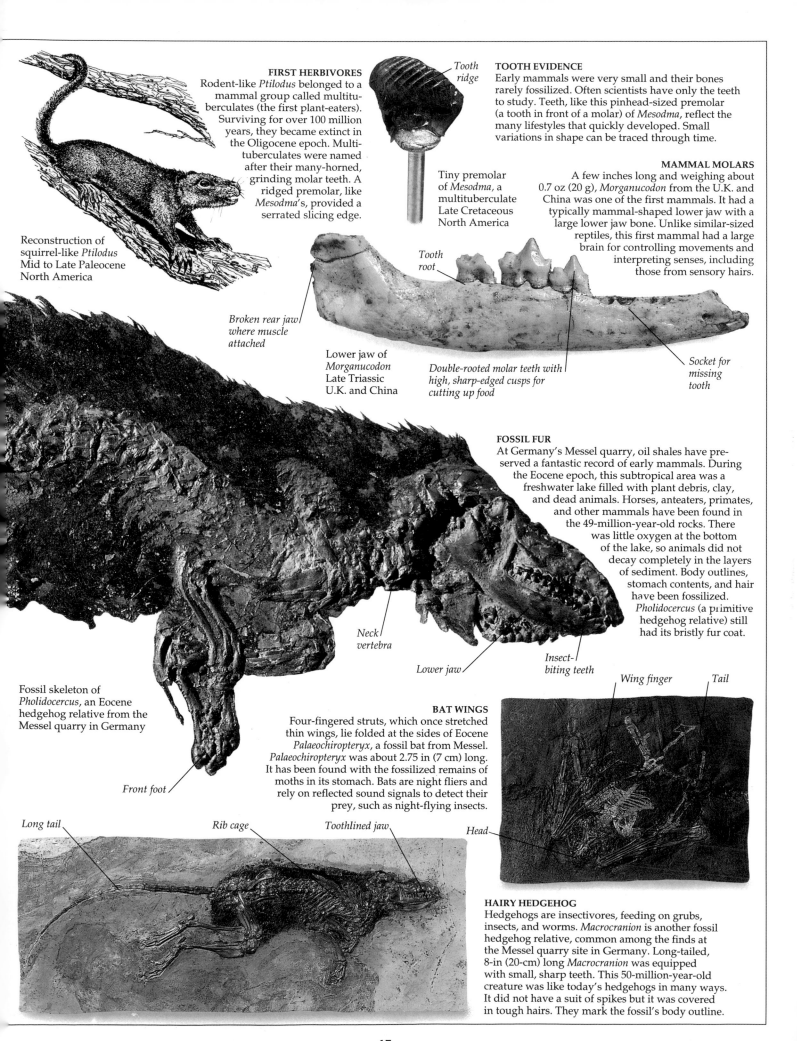

FIRST HERBIVORES
Rodent-like *Ptilodus* belonged to a mammal group called multituberculates (the first plant-eaters). Surviving for over 100 million years, they became extinct in the Oligocene epoch. Multituberculates were named after their many-horned, grinding molar teeth. A ridged premolar, like *Mesodma*'s, provided a serrated slicing edge.

Reconstruction of squirrel-like *Ptilodus* Mid to Late Paleocene North America

Tooth ridge

Tiny premolar of *Mesodma*, a multituberculate Late Cretaceous North America

TOOTH EVIDENCE
Early mammals were very small and their bones rarely fossilized. Often scientists have only the teeth to study. Teeth, like this pinhead-sized premolar (a tooth in front of a molar) of *Mesodma*, reflect the many lifestyles that quickly developed. Small variations in shape can be traced through time.

MAMMAL MOLARS
A few inches long and weighing about 0.7 oz (20 g), *Morganucodon* from the U.K. and China was one of the first mammals. It had a typically mammal-shaped lower jaw with a large lower jaw bone. Unlike similar-sized reptiles, this first mammal had a large brain for controlling movements and interpreting senses, including those from sensory hairs.

Tooth root

Broken rear jaw where muscle attached

Lower jaw of *Morganucodon* Late Triassic U.K. and China

Double-rooted molar teeth with high, sharp-edged cusps for cutting up food

Socket for missing tooth

FOSSIL FUR
At Germany's Messel quarry, oil shales have preserved a fantastic record of early mammals. During the Eocene epoch, this subtropical area was a freshwater lake filled with plant debris, clay, and dead animals. Horses, anteaters, primates, and other mammals have been found in the 49-million-year-old rocks. There was little oxygen at the bottom of the lake, so animals did not decay completely in the layers of sediment. Body outlines, stomach contents, and hair have been fossilized. *Pholidocercus* (a primitive hedgehog relative) still had its bristly fur coat.

Neck vertebra

Lower jaw

Insect-biting teeth

Fossil skeleton of *Pholidocercus*, an Eocene hedgehog relative from the Messel quarry in Germany

Front foot

BAT WINGS
Four-fingered struts, which once stretched thin wings, lie folded at the sides of Eocene *Palaeochiropteryx*, a fossil bat from Messel. *Palaeochiropteryx* was about 2.75 in (7 cm) long. It has been found with the fossilized remains of moths in its stomach. Bats are night fliers and rely on reflected sound signals to detect their prey, such as night-flying insects.

Wing finger

Tail

Head

Long tail

Rib cage

Toothlined jaw

HAIRY HEDGEHOG
Hedgehogs are insectivores, feeding on grubs, insects, and worms. *Macrocranion* is another fossil hedgehog relative, common among the finds at the Messel quarry site in Germany. Long-tailed, 8-in (20-cm) long *Macrocranion* was equipped with small, sharp teeth. This 50-million-year-old creature was like today's hedgehogs in many ways. It did not have a suit of spikes but it was covered in tough hairs. They mark the fossil's body outline.

Mammal variety

Mammals adapted to a wide variety of lifestyles. Over millions of years, the early shrew-like mammals evolved into ferocious hunters, bearlike browsers, and hoofed herbivores. Some mammals, such as whales, turned to a life in the sea. Others, such as bats, even managed to fly. Unlike reptiles, mammals did not rely on the Sun's heat for energy. As long as they had enough food, they were not tied to one climate or area. Mammal teeth provide us with clues to their diet. Carnivorous mammals have blade-like teeth for slicing and chewing meat; plant-eaters have ridged, flat teeth that can grind up plants. There are few places where mammals have not found a home.

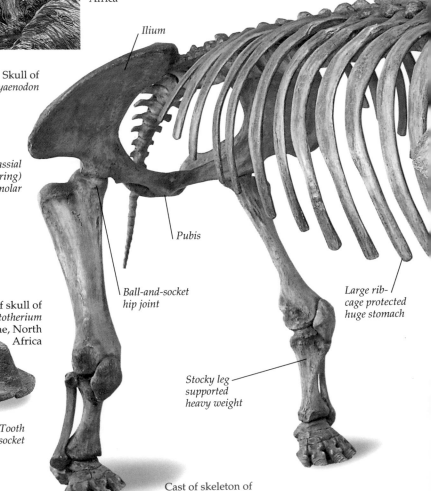

Top view of crest of *Prolibytherium* Early Miocene Libya

Attachment to skull

WINGED HORNS
The winged head-plate of *Prolibytherium* (a small, slim, deerlike member of the giraffe family) looks little like the ossicones (head growths) of living giraffes. Giraffes were a diverse group of animals during the Miocene and Pliocene epochs in Africa and Asia.

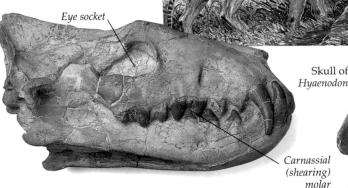

Large-headed *Hyaenodon* standing by dead prey Early Tertiary North America, Europe, and Africa

Eye socket

Skull of *Hyaenodon*

Carnassial (shearing) molar

Ilium

SMALL-BRAINED CARNIVORES
Now extinct, creodonts were the top carnivores in the early Tertiary period. *Hyaenodon* was a creodont which sometimes reached the size of today's living hyenas. Its carnassial teeth were adapted for shearing flesh, and its long canines for stabbing. Creodonts had relatively large heads compared with modern carnivores – and remarkably small brains.

Pubis

Ball-and-socket hip joint

Large rib-cage protected huge stomach

Crest for muscle attachment

Cast of skull of *Megistotherium* Miocene, North Africa

Stocky leg supported heavy weight

Tooth socket

SAVAGE SKULL
Megistotherium was an enormous creodont that lived about 20 mya in what is now the Sahara Desert. *Megistotherium*'s skull was 26 in (65 cm) long, twice the size of a lion's skull. The wide cheekbones and high-crested skull show that *Megistotherium* had enormous muscles to power its savage bite.

Cast of skeleton of *Arsinoitherium* Oligocene Egypt

EARLY WHALE

Early whale *Basilosaurus* had a long body and small head. Front limbs were modified for swimming, rear limbs had nearly disappeared. The long snout had nostrils on its upper surface. Later whales breathed through special nostrils at the top of the head.

Model of 66-ft (20-m) long *Basilosaurus* Eocene, Atlantic coast of the U.S.

Front limb was a flipper

Small trace of rear limb

Long tail

Eye socket

Front ossicone (head growth)

Nasal bone

Short-necked *Sivatherium* looked more like a giant elk than a giraffe

Skull of *Sivatherium giganteum*, Pliocene, India

TOUCHY NOSE

Sivatherium was a gigantic giraffid that lived in India and Africa during the Pliocene epoch. Only the front pair of ossicones are preserved in this fossil skull; a larger, heavier pair flared out from the back of the head. The strangely shaped nose bones suggest that *Sivatherium* may have had a prehensile (grasping) nose, as found in living tapirs.

Carnassial tooth

Neural spine on vertebra

Strong neck vertebra

Skull of *Daphoenus* Oligocene, U.S.

BEAR DOG

Bear dogs, the amphicyonids, were an important group of carnivore mammals ranging from the late Eocene to the Pliocene epochs. *Daphoenus* was one of the smaller amphicyonids, with a length of 3.3 ft (1 m). Many amphicyonids were large, stocky animals with bearlike bodies and doglike skulls.

Movable joint in neck

Small horn over eyebrow

A pair of long-tailed, lightly built *Daphoenus* bear dogs on the prowl for prey

Deep-set eye socket

Bony horn with furrows left by blood vessels

Broad, chisel-edged molar for grinding plant food

Nostril

POINTED HEAD

Horns of bone rather than tightly packed hair distinguish *Arsinoitherium* from living rhinoceroses. From the Oligocene epoch of Egypt, *Arsinoitherium* also differed from rhinos in having a complete set of incisor teeth. The molar teeth had large, crescent-shaped cutting edges, a feature found today in mammals whose diet consists mainly of leaves. Five fat toes on the end of each short, stocky limb carried its vast weight. *Arsinoitherium* belonged to a group of mammals called embrithopods, which became extinct in the Oligocene epoch.

Blocky, five-toed foot held up massive body

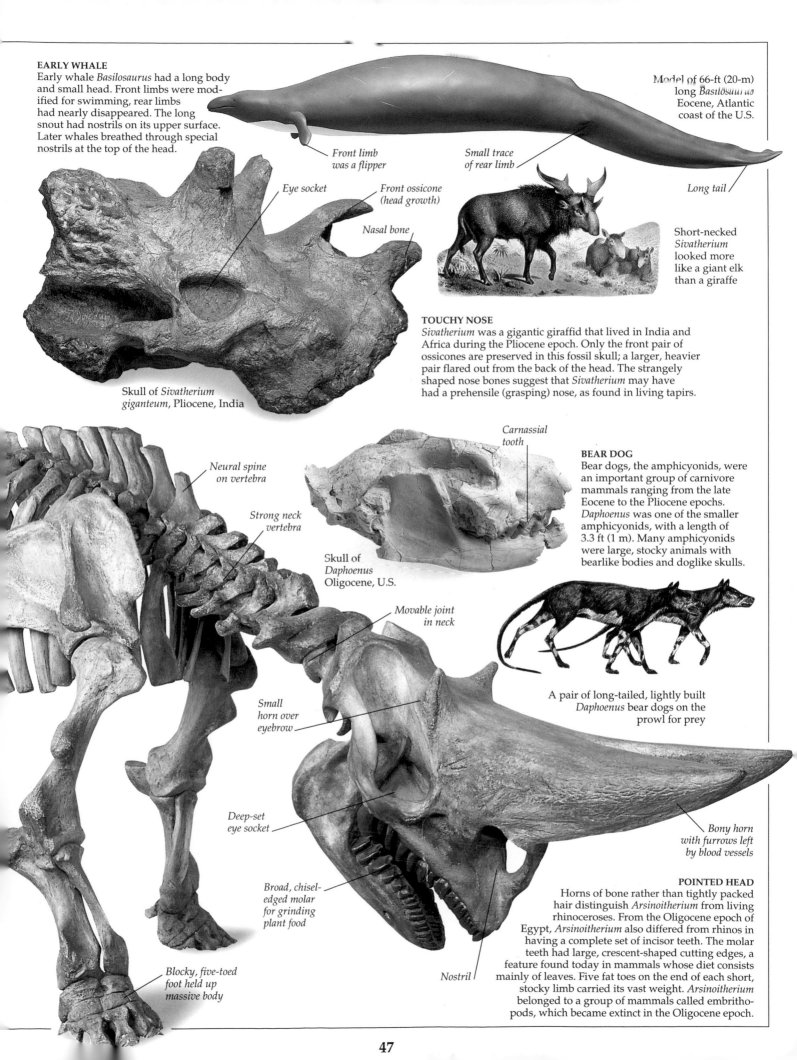

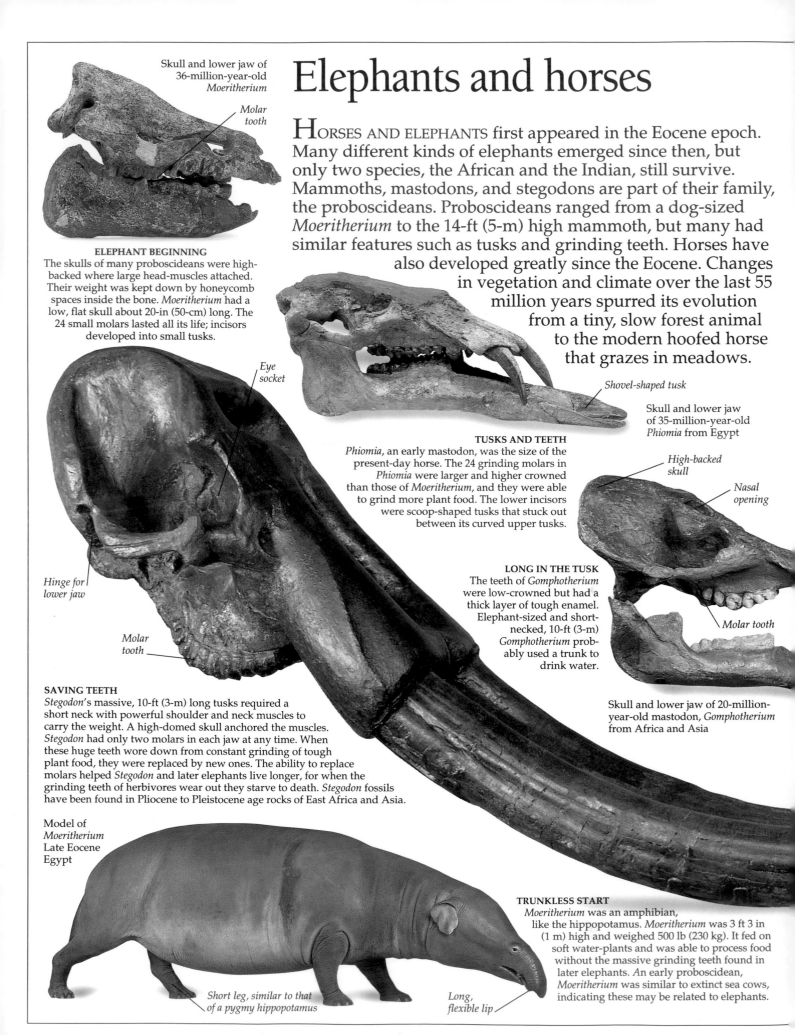

Elephants and horses

HORSES AND ELEPHANTS first appeared in the Eocene epoch. Many different kinds of elephants emerged since then, but only two species, the African and the Indian, still survive. Mammoths, mastodons, and stegodons are part of their family, the proboscideans. Proboscideans ranged from a dog-sized *Moeritherium* to the 14-ft (5-m) high mammoth, but many had similar features such as tusks and grinding teeth. Horses have also developed greatly since the Eocene. Changes in vegetation and climate over the last 55 million years spurred its evolution from a tiny, slow forest animal to the modern hoofed horse that grazes in meadows.

Skull and lower jaw of 36-million-year-old Moeritherium

Molar tooth

ELEPHANT BEGINNING
The skulls of many proboscideans were high-backed where large head-muscles attached. Their weight was kept down by honeycomb spaces inside the bone. *Moeritherium* had a low, flat skull about 20-in (50-cm) long. The 24 small molars lasted all its life; incisors developed into small tusks.

Eye socket

Hinge for lower jaw

Molar tooth

SAVING TEETH
Stegodon's massive, 10-ft (3-m) long tusks required a short neck with powerful shoulder and neck muscles to carry the weight. A high-domed skull anchored the muscles. *Stegodon* had only two molars in each jaw at any time. When these huge teeth wore down from constant grinding of tough plant food, they were replaced by new ones. The ability to replace molars helped *Stegodon* and later elephants live longer, for when the grinding teeth of herbivores wear out they starve to death. *Stegodon* fossils have been found in Pliocene to Pleistocene age rocks of East Africa and Asia.

Model of
Moeritherium
Late Eocene
Egypt

Short leg, similar to that of a pygmy hippopotamus

Shovel-shaped tusk

Skull and lower jaw of 35-million-year-old *Phiomia* from Egypt

TUSKS AND TEETH
Phiomia, an early mastodon, was the size of the present-day horse. The 24 grinding molars in *Phiomia* were larger and higher crowned than those of *Moeritherium*, and they were able to grind more plant food. The lower incisors were scoop-shaped tusks that stuck out between its curved upper tusks.

High-backed skull

Nasal opening

LONG IN THE TUSK
The teeth of *Gomphotherium* were low-crowned but had a thick layer of tough enamel. Elephant-sized and short-necked, 10-ft (3-m) *Gomphotherium* prob-ably used a trunk to drink water.

Molar tooth

Skull and lower jaw of 20-million-year-old mastodon, *Gomphotherium* from Africa and Asia

TRUNKLESS START
Moeritherium was an amphibian, like the hippopotamus. *Moeritherium* was 3 ft 3 in (1 m) high and weighed 500 lb (230 kg). It fed on soft water-plants and was able to process food without the massive grinding teeth found in later elephants. An early proboscidean, *Moeritherium* was similar to extinct sea cows, indicating these may be related to elephants.

Long, flexible lip

Hooves and teeth

Horses have gone through their own distinct set of changes since they first appeared in the Eocene forests of 55 mya. Smaller than a domestic cat, the first horse, *Hyracotherium*, was a browser, feeding on seeds, fruits, and young leaves. It was not a fast animal, but padded on four-toed feet around forests. Over the next 55 million years, horses developed longer teeth to deal with tough, abrasive grasses.

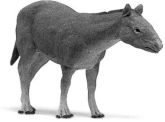

FIRST HORSE
Inhabiting North America and Europe, *Hyracotherium* was the first in a long line of horses, from forest browsers to plains grazers.

Front foot of early four-toed horse, *Hyracotherium* (55 mya)

TOE LOSS
In the progression to fewer toes, toe three became larger and longer until it became as broad as the horse's leg. Side toes were reduced and eventually lost in modern horses. Three-toed *Hipparion* placed most of its weight on its large middle toe. It probably managed a running walk of up to 9 mph (15 km/h).

THREE-TOED HORSE
Three-toed *Merychippus* marked a major step in horse evolution. It was the first of the grazing horses and lived about 20 mya in North America.

Left hind foot of three-toed *Hipparion*, which lived from 15 mya in Europe, Asia, Africa, and North America

Feet, with fewer toes, geared to fast running on open plains over 55 million years

Low-crowned molar

Bottom view of *Hyracotherium* skull

High-crowned tooth

Incisor

Bottom view of *Hipparion* skull

GRASS GRINDERS
During the early history of the horse, changes in climate reduced forests, and about 23 mya, grass plains spread over wide areas. Horses, such as *Hipparion*, adapted to a grazing way of life. They developed long, heavily ridged teeth. The ridges on the grinding surfaces were made of hard, tough tooth enamel. Like coarse files, the ridged tooth rows could grind the tough cellulose of grass plants.

Tusks of Stegodon were 10 ft (3 m) long

Skull and tusks of Gomphotherium were up to 6 ft 7 in (2 m) long

Skull and tusks of five-million-year-old *Stegodon ganesa* from India

TODAY'S ELEPHANTS
Elephants related to those living today first appeared about 7 mya. Today's elephants have only one large tooth in each jaw at a time. The back teeth slowly grow forward as the teeth in front wear down. New teeth replace old ones throughout life. Elephants can weigh close to 12 tons and must eat an enormous amount each day to survive.

SHORT TUSKS
Short-trunked *Phiomia* lived in North Africa about 32 mya. No more than 6 ft 7 in (2 m) tall, *Phiomia* was clearly elephant-like although its trunk, tusks, and body were still small. Cusped teeth had to grind through large volumes of plant food, wearing down to a flat grinding surface as they became older.

Short tusk used for rooting up plants

Model of *Phiomia* Early Oligocene, Egypt

Mammal islands

Engraving of *Diprotodon* from Australia

WHEN PANGAEA, the Mesozoic super-continent, began to break up in the Jurassic period, each rift and seaway was an effective barrier to some groups of animals. South America and Australia became huge continental islands separated by sea barriers from other lands, and their mammal populations developed in isolation. South America and North America broke apart about 55 million years ago, at a time when marsupial (pouched) mammals were as widespread as placentals. For more than 50 million years, South America developed its own distinct marsupial and placental animal population, cut off from the rest of the world. About 3 million years ago, this mammal isolation ended when a bridge of land (the isthmus of Panama) developed, providing a two-way migration route between North and South America. The marsupials of Australia had come from Antarctica when it was much warmer than today. Australia broke away about 35 million years ago and its mammal isolation has been more complete than the Americas'.

PLEISTOCENE SCENE
Pleistocene Australia was a land of wondrous marsupial animals. Some became extinct only recently. Koalas, kangaroos, and the strangely primitive, spiky echidna, an egg-laying mammal (bottom right), have survived to the present. However, wombat-like *Diprotodon*, lion-like *Thylacoleo*, and wolf-like *Thylacinus* have become extinct, although the latter only recently.

High nasal bone

Attachment area for neck muscles

Flat grinding tooth

Curved incisor

Wide pelvis for leg muscles

Marsupial bone supported pouch

OUTSIZE AUSTRALIAN
An outsize marsupial, *Diprotodon* lived during the Pleistocene period of Australia. About 10 ft (3 m) long, *Diprotodon* was one of the largest plant-eating marsupials around. It had broad, plant-grinding molar teeth and huge incisors, like those of wombats. Short-limbed, stocky *Diprotodon* may have been still alive when the Aboriginals first inhabited Australia.

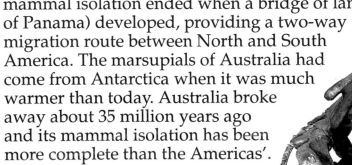

Pelvis

Marsupial bone

Skeleton of *Sthenurus tindalei*, Pleistocene Australia

Large heel

Skeleton of *Diprotodon*, Pleistocene Australia

Flat foot

Fossil skeleton of *Thylacoleo* Pleistocene Australia

Single long middle toe

Long fingered hand

HOOFED HOPPER
Fast-moving hoppers, kangaroos may be up to 6 ft 6 in (2 m) tall, but some early kangaroos must have been taller than 10 ft (3 m) when alive. Short-jawed *Sthenurus* was a large Pliocene kangaroo, but unlike other kangaroos its large rear foot had a single toe that ended in a hoof-shaped bone. It had a modified shoulder blade that helped it reach high above its head to eat young leaves at the tips of branches.

MAMMAL MIMIC
Wolflike *Thylacinus* was not the only mammal mimic. Marsupial *Thylacoleo* looked like a lion with its sharp, stabbing and tearing incisor teeth and its slicing carnassial back teeth. *Thylacoleo* hunted Australia's Pleistocene marsupial herbivores, such as *Diprotodon* and kangaroos.

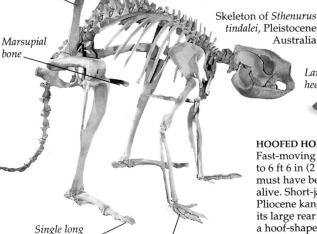

Distinctive stripes on back and tail

Reconstruction of *Thylacinus*, the "Tasmanian wolf" from the Miocene period of Australia; it only recently became extinct

TASMANIAN WOLF
Some marsupials were fearsome carnivores. *Thylacinus*, although similar in general appearance to the placental wolf of the northern hemisphere, was very definitely a marsupial. The body shape and skulls are very much alike, and *Thylacinus* probably lived the same kind of pack-hunting life as its unrelated northern double. Known as the "Tasmanian wolf," *Thylacinus* became extinct in 1934.

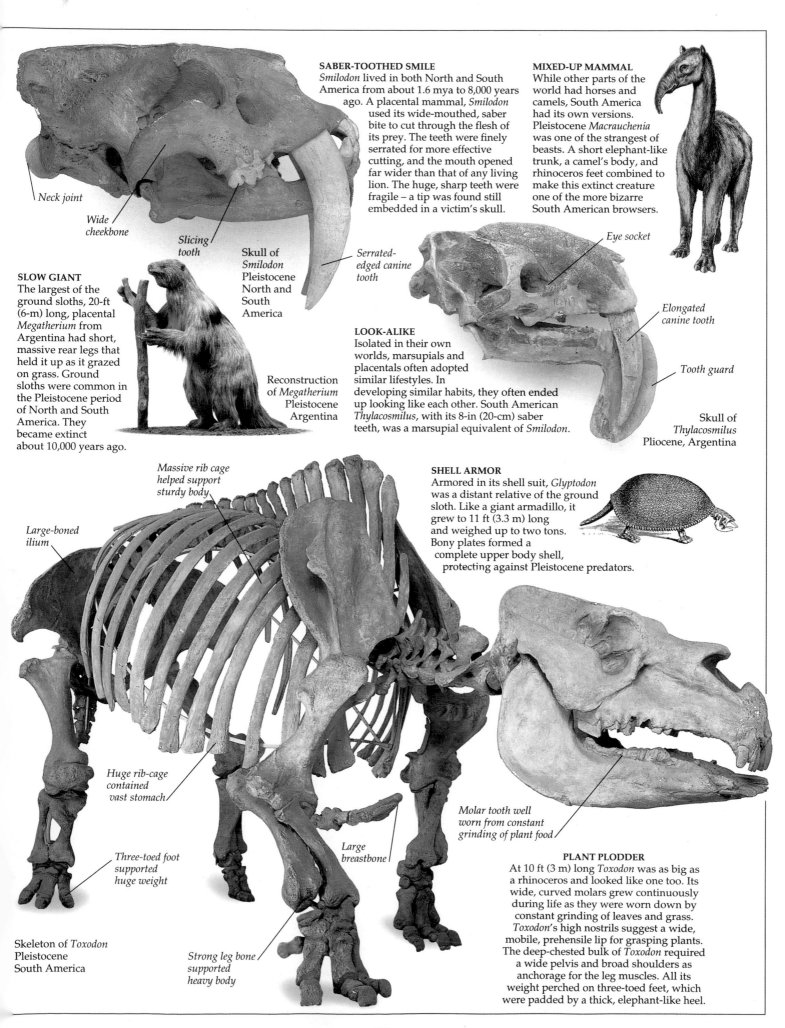

SABER-TOOTHED SMILE

Smilodon lived in both North and South America from about 1.6 mya to 8,000 years ago. A placental mammal, *Smilodon* used its wide-mouthed, saber bite to cut through the flesh of its prey. The teeth were finely serrated for more effective cutting, and the mouth opened far wider than that of any living lion. The huge, sharp teeth were fragile – a tip was found still embedded in a victim's skull.

Neck joint

Wide cheekbone

Slicing tooth

Skull of *Smilodon* Pleistocene North and South America

Serrated-edged canine tooth

MIXED-UP MAMMAL

While other parts of the world had horses and camels, South America had its own versions. Pleistocene *Macrauchenia* was one of the strangest of beasts. A short elephant-like trunk, a camel's body, and rhinoceros feet combined to make this extinct creature one of the more bizarre South American browsers.

Eye socket

Elongated canine tooth

Tooth guard

Skull of *Thylacosmilus* Pliocene, Argentina

SLOW GIANT

The largest of the ground sloths, 20-ft (6-m) long, placental *Megatherium* from Argentina had short, massive rear legs that held it up as it grazed on grass. Ground sloths were common in the Pleistocene period of North and South America. They became extinct about 10,000 years ago.

Reconstruction of *Megatherium* Pleistocene Argentina

LOOK-ALIKE

Isolated in their own worlds, marsupials and placentals often adopted similar lifestyles. In developing similar habits, they often ended up looking like each other. South American *Thylacosmilus*, with its 8-in (20-cm) saber teeth, was a marsupial equivalent of *Smilodon*.

SHELL ARMOR

Armored in its shell suit, *Glyptodon* was a distant relative of the ground sloth. Like a giant armadillo, it grew to 11 ft (3.3 m) long and weighed up to two tons. Bony plates formed a complete upper body shell, protecting against Pleistocene predators.

Massive rib cage helped support sturdy body

Large-boned ilium

Huge rib-cage contained vast stomach

Three-toed foot supported huge weight

Large breastbone

Strong leg bone supported heavy body

Skeleton of *Toxodon* Pleistocene South America

Molar tooth well worn from constant grinding of plant food

PLANT PLODDER

At 10 ft (3 m) long *Toxodon* was as big as a rhinoceros and looked like one too. Its wide, curved molars grew continuously during life as they were worn down by constant grinding of leaves and grass. *Toxodon*'s high nostrils suggest a wide, mobile, prehensile lip for grasping plants. The deep-chested bulk of *Toxodon* required a wide pelvis and broad shoulders as anchorage for the leg muscles. All its weight perched on three-toed feet, which were padded by a thick, elephant-like heel.

Birth and growth

LIFE HAS SURVIVED ON EARTH for more than three billion years. From the first, it found a way to reproduce itself. Prokaryote, blue-green algae simply divided their cells in two, each exactly like the other. Sexual reproduction increased the variety of life. Male and female cells combined to produce an offspring different from both parents, increasing the chances of further change. Some animals, such as fish and amphibians, laid their fragile eggs in water, but the first reptile produced a new, amniotic egg. Protected by a tough, waterproof eggshell, the embryo developed in its own watery environment enclosed within the amnion (embryo sac). Reptiles were freed to live on land, though some developed as marine animals giving birth at sea. Mammals, which also have amniotic eggs, gave them further protection by keeping the eggs and developing young inside their bodies until born.

TADPOLE TALE
Amphibian eggs are covered in a thick layer of protective albumen (the white of the egg). In water, the albumen becomes sticky, holding the eggs together as a mass. The frogs then pass through a tadpole stage. This rare fossil never got any further.

BRANCHING OUT
Set in their stony skeletons, the coral polyps produce side branches by building. Each new branch adds to the coral colony. At intervals, the polyps also eject male and female cells which combine to produce free-swimming coral animals. Survivors settle and grow into new coral colonies.

Coral released egg into water

Thecosmilia trichotoma (colony of coral with separate branches), Late Jurassic Germany

Rejuvenating corallite branch increases colony

Red shows one embryo (of six) that died with adult female

Stenopterygius quadriscissus Early Jurassic, Germany

Maiasaura eggs were laid in a scooped-out hollow in sand and covered with vegetation, as shown in this model

FOSSIL BIRTH
The fish-shaped ichthyosaurs descended from land reptiles that had returned to the sea. They did not lay eggs but gave birth to live young in the water. Ichthyosaur fossils have been found with the young still inside the body. This *Stenopterygius* probably died from some difficulty during birth.

INSECT TRAP
Trapped in the sticky resin of a tree, these mating flies were fossilized in golden amber. At some stage, insects "invented" metamorphosis, growth by changing form. Instead of going from an egg to a small version of an adult that grows in size, insects change from egg to larva (like a caterpillar) to pupa to adult.

Aepyornis lived during the Pleistocene epoch on Madagascar and laid its eggs in sand dunes

RECORD STORE

Like many other mollusks, an ammonoid records the history of its growth in its coiled shell, called an ammonite. At the center of the coil is the protoconch (first small shelly cover). As it grew outward, the tentacled ammonoid occupied a series of chambers. Each chamber, larger than the previous one, was a new stage in its growth. The chambers, coiling out from the center, were filled with gas and kept the animal buoyant in the sea.

ELEPHANT-BIRD EGG

It is not surprising that giant birds should lay giant eggs. The 10-ft (3-m) tall, flightless *Aepyornis* holds the record. Weighing about 1,000 lb (450 kg), *Aepyornis* laid eggs 170 times larger than those of hens. However, larger eggs need thicker shells, which limits their size. If the shell was too thick, the chick could not break out.

PALM FRUIT

Plants developed many ways to spread their seeds. When a tasty fruit was eaten by an animal, the tough seeds survived intact and were excreted, often far from their source. The coconut-like fruit of *Nipa*, a stemless palm, enclosed seeds up to 3 in (8 cm) long. The fruit was probably either carried away by water or split open in *Nipa*'s tropical habitat.

Woody outer layer enclosed large seeds in fibrous inner layer

Fruit of *Nipa burtinii*, Mid Eocene, Belgium

Oxynoticeras oxynotum, Early Jurassic, U.K.

Chamber filled with crystals during fossilization

DINOSAUR PLAYPEN

Dinosaur eggs were first recognized in 1922, when fossilized nests of *Protoceratops* were discovered in the Gobi Desert. Since then many more dinosaur eggs have been discovered. Some of the most exciting discoveries were made in Montana, where nesting sites of *Maiasaura* were found in 1978. *Maiasaura* ("good mother lizard") not only laid its eggs in scooped-out nests but also cared for the young, 14-in (36-cm) hatchlings until they were large enough to look after themselves.

Apes and ancestors

We HUMANS ARE UNIQUE in our upright walk, spoken language, advanced use of tools and fire, and level of intelligence. However, evidence from our body shape points to the sociable, tool-using chimpanzees as our closest living relatives. Although it may offend some people, scientific evidence places humans, gorillas, and chimpanzees in the same group – Hominoidea. On a wider scale, humans are members of the Primates group, along with 200 other species of monkeys, apes, and prosimians (a group of animals including lemurs and bush babies). This does not mean that gorillas and chimpanzees were our ancestors but that at some point in the past, humans and apes had a common ancestor. Since then apes have developed along their own lines and humans along another. Fossils of primates help to pinpoint the junctions where lines branched, where groups split and headed off in new and different directions.

No eyebrow ridge

Shallow eye socket

AEGYPTOPITHECUS
Discovered in the desert of Fayum in Egypt, tree-living *Aegyptopithecus* lived about 32 mya when the area was a tropical rainforest. Although monkey-like, this small primate was distinct from both apes and monkeys and may have been ancestral to both.

PROCONSUL
Named in 1933, *Proconsul africanus* combined the skeletal features of modern monkeys and apes. A representative of the earliest apes, *Proconsul* had the hands, arms, and long body of tree-swinging monkeys but it also had the skull, shoulders, and elbows of knuckle-walking apes. There were several species of *Proconsul* of very different sizes. These first apes of 18 mya fed in the woods and forests of what is now Kenya.

Partial skull of *Sivapithecus indicus* (10.5–8 mya) Pakistan

Long face curves down to projecting mouth

Thickened ridge on lower side of jaw

Small lateral incisor, as in today's orangutan

ORANG ANCESTOR
Making links between living hominoids and fossils is difficult. It is only the orangutan that can be linked reasonably well to an early ape. *Sivapithecus*, found in Miocene rocks of 13–8 mya, displays the long face, projecting mouth, and flattened cheeks that help identify this orang-utan line of apes. More distant from humans than the gorilla and chimpanzee, *Sivapithecus* was an early branch off the line that went on to produce the African apes and humans. Named after the Indian god Siva and found in India, Pakistan, and Turkey, *Sivapithecus* is the earliest hominoid outside Africa.

Flat, thick tooth enamel similar to that of an orangutan

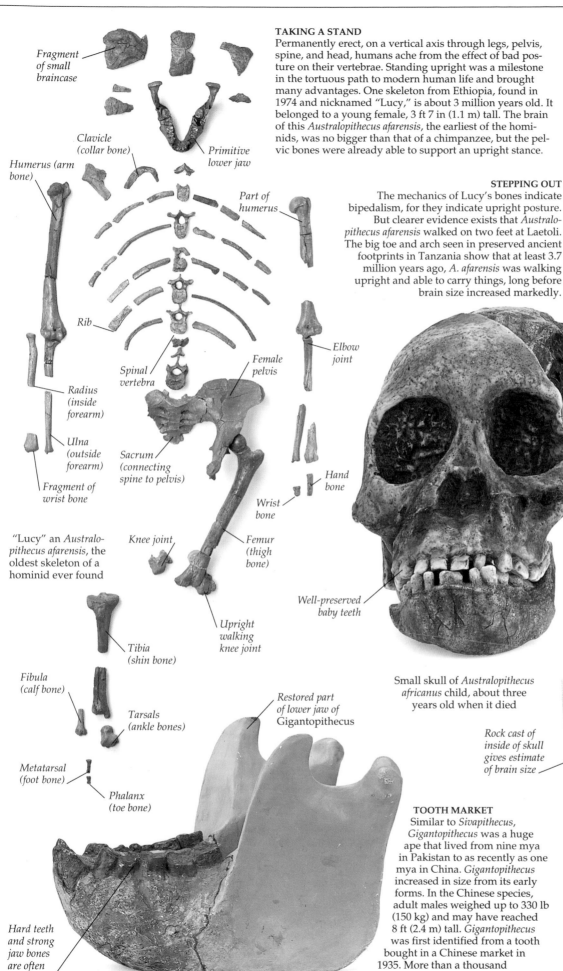

TAKING A STAND

Permanently erect, on a vertical axis through legs, pelvis, spine, and head, humans ache from the effect of bad posture on their vertebrae. Standing upright was a milestone in the tortuous path to modern human life and brought many advantages. One skeleton from Ethiopia, found in 1974 and nicknamed "Lucy," is about 3 million years old. It belonged to a young female, 3 ft 7 in (1.1 m) tall. The brain of this *Australopithecus afarensis*, the earliest of the hominids, was no bigger than that of a chimpanzee, but the pelvic bones were already able to support an upright stance.

STEPPING OUT

The mechanics of Lucy's bones indicate bipedalism, for they indicate upright posture. But clearer evidence exists that *Australopithecus afarensis* walked on two feet at Laetoli. The big toe and arch seen in preserved ancient footprints in Tanzania show that at least 3.7 million years ago, *A. afarensis* was walking upright and able to carry things, long before brain size increased markedly.

Mary Leakey (pp. 56–57) examining two trails of footprints at Laetoli in northern Tanzania

Fragment of small braincase

Clavicle (collar bone)

Primitive lower jaw

Humerus (arm bone)

Part of humerus

Rib

Elbow joint

Spinal vertebra

Female pelvis

Radius (inside forearm)

Ulna (outside forearm)

Sacrum (connecting spine to pelvis)

Hand bone

Fragment of wrist bone

Wrist bone

Knee joint

Femur (thigh bone)

"Lucy" an *Australopithecus afarensis*, the oldest skeleton of a hominid ever found

Upright walking knee joint

Tibia (shin bone)

Fibula (calf bone)

Tarsals (ankle bones)

Metatarsal (foot bone)

Phalanx (toe bone)

Restored part of lower jaw of Gigantopithecus

Hard teeth and strong jaw bones are often fossilized

FOSSIL CHILD

Australopithecus afarensis is the oldest known hominid, advanced in its ability to walk upright but still with apelike features. The mixture of ape and human features was recognized in the first *Australopithecus* discovered, the Taung skull. This small *A. africanus* skull (2.3 to 2 million years old) is more human in its flatter face, small canine teeth, and upright head than *A. afarensis*. *Australopithecus africanus* adults grew to 4 ft (1.2 m) in height, and weighed about 77 lb (35 kg).

Well-preserved baby teeth

Small skull of *Australopithecus africanus* child, about three years old when it died

Rock cast of inside of skull gives estimate of brain size

TOOTH MARKET

Similar to *Sivapithecus*, *Gigantopithecus* was a huge ape that lived from nine mya in Pakistan to as recently as one mya in China. *Gigantopithecus* increased in size from its early forms. In the Chinese species, adult males weighed up to 330 lb (150 kg) and may have reached 8 ft (2.4 m) tall. *Gigantopithecus* was first identified from a tooth bought in a Chinese market in 1935. More than a thousand specimens have been found since.

TAUNG DISCOVERER

Named *Australopithecus africanus* (southern ape from Africa) by Raymond Dart in 1925, the Taung skull had been blown from rock during quarry blasting. A professor of Anatomy at Witwatersrand University in South Africa, Dart caused a sensation by interpreting this fossil as an intermediate between apes and humans. It took 20 years for many scientists to accept that it was in Africa, not Asia, that the human animal emerged and that a large brain did not evolve before an upright stance.

Early humans

THE ORIGIN OF MODERN HUMANS (*Homo sapiens*) is a subject of serious debate among scientists. The arguments are heated because discoveries of fossil humans are relatively rare, many having been made only since the early 1960s. Every detail revealed is important, as changes in humans have occurred in a relatively short time. Modern humans may have appeared as recently as 100,000 years ago. In some places they lived alongside Neanderthals for over 40,000 years. Neanderthals changed little, remaining a distinct people until they died out about 30,000 years ago. Modern humans seem to have appeared first in Africa, and then spread through the world.

OLDUVAI GORGE
In Tanzania's Serengeti Plain, Olduvai Gorge with its layers of volcanic and sedimentary rock (1.9 to 0.1 million years old) is the site where the first known human, *Homo habilis*, and the later *Homo erectus*, were discovered. British paleontologists Louis Leakey (1903–1972) and his wife, Mary (b. 1913), discovered the skull of a 1.8 million-year-old hominid in 1964.

Body completely covered with hair

Cranium of Homo habilis

Teeth like those of A. africanus

HANDY HUMAN
The discovery of *Homo habilis* at Olduvai and at other African sites prompted a debate about its *Australopithecus*-like flat-sided face and tree-climbing ability compared with their more human-sized brain capacity and precise gripping hand.

Stone hammer for smashing bones or shaping tools

TOOL MAKER
Some scientists believe that the differences between *Australopithecus africanus* (pp. 54–55) and later *Homo erectus* are so small that intermediate *H. habilis* remains are really *A. africanus* or *H. erectus*. One robust kind of *Australopithecus* lived at Olduvai at the same time as *H. habilis*. The probable presence of larger-brained *H. habilis* is seen in the rough stone tools for scraping, cutting, and chopping found at Olduvai. The oldest tools of these kind (2.5 million years old) come from sites farther north in Ethiopia. *H. habilis* could have used them to cut meat and break open bones and hard fruits.

Model of Homo habilis, an upright walker with a large brain

Homo habilis may have had a tool kit of differently shaped stone tools

Cranium of *Homo erectus* from Koobi Fora in northern Kenya

Low skull cap with jutting brow ridge

Cranium of *Homo neanderthalensis* from Neander Valley in Germany

Skull of male Neanderthal found at La Ferrassie

Brain as large as modern human's

Low forehead

Heavy, projecting brow ridge

Wide-set eyes

Broad-based skull

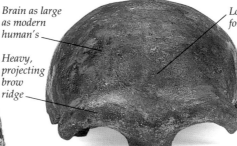

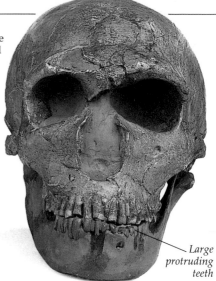

Large protruding teeth

THICK SKULL
Discovered in 1856, the thick skull cap and some limb bones were the first Neanderthal fossils to be recognized. Named after the site of their discovery, the Neander Valley near Dusseldorf in Germany, Neanderthals probably appeared over 200,000 years ago. They are well known from remains between 70,000 and 35,000 years old.

LARGE BRAIN
Neanderthals had a flat-topped, low skull with prominent bony eyebrow ridges. Their large face was marked by a huge nose, receding cheek bones, and large projecting teeth. The lower jaw often had no bony chin, as in this specimen from La Ferrassie Cave in southwest France. On average, a Neanderthal brain was larger (88.5 cu in, 1,450 cc) than a modern human's, but this 50,000-year-old male skull had a brain capacity of 97.5 cu in (1,600 cc). Neanderthals lived alongside modern humans in some places but remained distinct. They are a dead end in the human story.

UPRIGHT HUMANS
Fossils of *Homo erectus* (upright human) were first found in Indonesia in 1890. Until the discovery of other fossils in Africa in 1954, *H. erectus* was only known from Indonesia and China. From the Leakeys' find in the 1960s, we now know that *H. erectus* first appeared in East Africa 1.8 mya and spread into Asia 1 mya. Larger brained than *H. habilis*, *H. erectus* flaked stones for handaxes and butchered meat.

Large lower jaw and teeth

Thick humerus (upper arm bone)

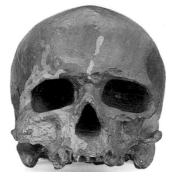

NEANDERTHAL PEOPLE
Neanderthals appeared over 200,000 years ago in Europe and western Asia. Stockily built, males were 5 ft 7 in (1.7 m) tall, females 5 ft 3 in (1.6 m). Neanderthals have a reputation for being stupid and primitive, but they were clever enough to survive in cold climates during Ice Ages. They made stone (but not antler or bone) tools.

TWO GROUPS
In Australia, the oldest human remains (30,000 years old), from Lake Mungo, are of slim, lightly-built people (above). Also present at the same time were people with heavier skulls. These two distinct groups may have been the result of more than one migration to this island continent.

HUMAN APPEARANCE
In Europe 30,000 years ago, modern humans dominated the cold, unfriendly landscape. These Cro-Magnons, named after their rock shelter in France, wore animal hides, made fine bone harpoons and delicate stone tools, and built huts or lived in caves. They were distinct from the disappearing Neanderthals, both in their culture and in their tall, slim appearance.

BURIAL RITES
Neanderthals may have been the first people to bury their dead. At Kebara Cave on Mount Carmel in Israel, 60,000 years ago, the large skeleton of a young man was deliberately buried without his skull. His massive jaw was covered up along with his skeleton after the flesh had rotted. Other Neanderthal burials may have been marked by flowers in the grave.

Cro-Magnon cranium

Ice ages

Panthera spelaea
Pleistocene
Europe

OVER MANY THOUSANDS OF YEARS, the Earth's path around the Sun alters so that parts of the Earth's surface are closer to or farther from the Sun and become warmer or cooler. During cold periods, winter snow may stay into summer. A small drop in global temperature, as little as 4°F (2°C) can disrupt the normal thawing of winter ice. Reflecting the Sun's rays, ice can lower the global temperature, and any volcanic dust thrown into the atmosphere can block the Sun and cool the Earth. Ice ages, with cold periods (glacials) that alternate with warm periods (interglacials), have been a recurring feature for at least two billion years. Over the past 1.6 million years, animal and plant life have had to deal with advancing and retreating ice sheets. Today, the Earth is in a warm, interglacial period.

Small ear cuts loss of body heat

NORTHERN LION
Lions live today only in Africa and India. In Europe, the cave lion (*Panthera spelaea*) died out 10,000 years ago. Its fossilized remains have been found in warm, interglacial deposits, where wide open grasslands let lions hunt for prey.

Hedera (Ivy)

Enormous ivory tusk curved up and inwards

MAMMOTH MONSTER
Of all Ice Age animals, the woolly mammoth, *Mammuthus primigenius*, is the most famous. About 10 ft (3 m) tall at the shoulders, mammoths had a long, shaggy coat of thick, dark hair and a layer of insulating fat. Great numbers of woolly mammoths ranged across the cold tundra regions of North America, Europe, and Asia. More than 500,000 tons of fossil tusks are estimated to be buried along 900 miles (1,500 km) of Siberian coastline. The tusks are collected for their ivory. Thought to have become extinct 10,000 years ago, recently found remains of smaller woolly mammoths show they were still alive 4,000 years ago.

PLANT THERMOMETER
Some plants thrive in warm climates while others can survive much lower temperatures. In northern Europe, ivy needs an average winter monthly temperature higher than 29°F (-1.5°C) to thrive. The Ice Age's warmer, interglacial periods can be identified by fossil ivy pollen from Pleistocene sediments.

DEEP FREEZE
Dima, a baby woolly mammoth, was just six months old when it died. Preserved in Arctic permafrost (permanently frozen soil), Dima remained frozen for 40,000 years until discovered by gold miners in 1977. Its carcass still had a coat of red hair.

Mollusk shell attached to root

Woolly mammoth tooth

DRY LAND
Dredged from the North Sea, this mammoth tooth with a mollusk shell stuck on its roots indicates a time when the sea floor was exposed as dry land. Ice caps trapped vast quantities of water and the sea level fell by as much as 330 ft (100 m). Ice Age mammals were able to cross land bridges from Alaska to Siberia, New Guinea to Australia, and the European mainland to Britain.

Thick woolly underhair

Hair covered trunk to keep out cold

Canine tooth

Molar tooth crushed food

Domed skull

Coelodonta antiquitatis
Late Pleistocene
Arctic

Mimomys savini
600,000 years ago

Arvicola cantiana
350,000 years ago

Arvicola terrestris
2,000 years ago

Ursus spelaeus
Pleistocene
Germany

PREHISTORIC PICTURES
Pushed south by the last great advance of ice 18,000 years ago, the European woolly rhino grazed on low grass alongside the woolly mammoths. Cave drawings produced by the Cro-Magnons (pp. 56–57) who lived in the Dordogne in France support the fossil evidence from European glacial deposits.

TINY TIMEPIECE
The crowns of the back teeth of water voles (above), the most common small mammals found in Pleistocene sediments, have a distinct pattern for each new species and are used for dating Ice Age deposits.

CAVE HOME
One of the most common carnivores during the Pleistocene Ice Ages, the cave bear was larger than today's European brown bear. Cave bear fossils are often found among the rock debris of cave floors, where the bears may have died while hibernating. One cave in Austria contained the remains of more than 30,000 cave bears, accumulated over many thousands of years.

Huge canine tusk

Stocky leg to support heavy body

Hippopotamus amphibius
Pleistocene
England

Four-toed foot evenly distributed body weight

HOT HIPPO
The last interglacial period in Britain, the Ipswichian of 120,000 years ago, brought "warm weather" animals north, such as spotted hyenas, bison, and this hippopotamus. The complete skeleton and other hippo fossils from southern Britain belong to *Hippopotamus amphibius*, which now lives only in Africa south of the Sahara Desert. As the climate cooled, hippos disappeared from Europe, replaced by "cold climate" animals.

Thick insulating coat, an adaptation to living in an extremely cold climate

Reconstructed *Mammuthus primigenius*, Pleistocene Arctic regions

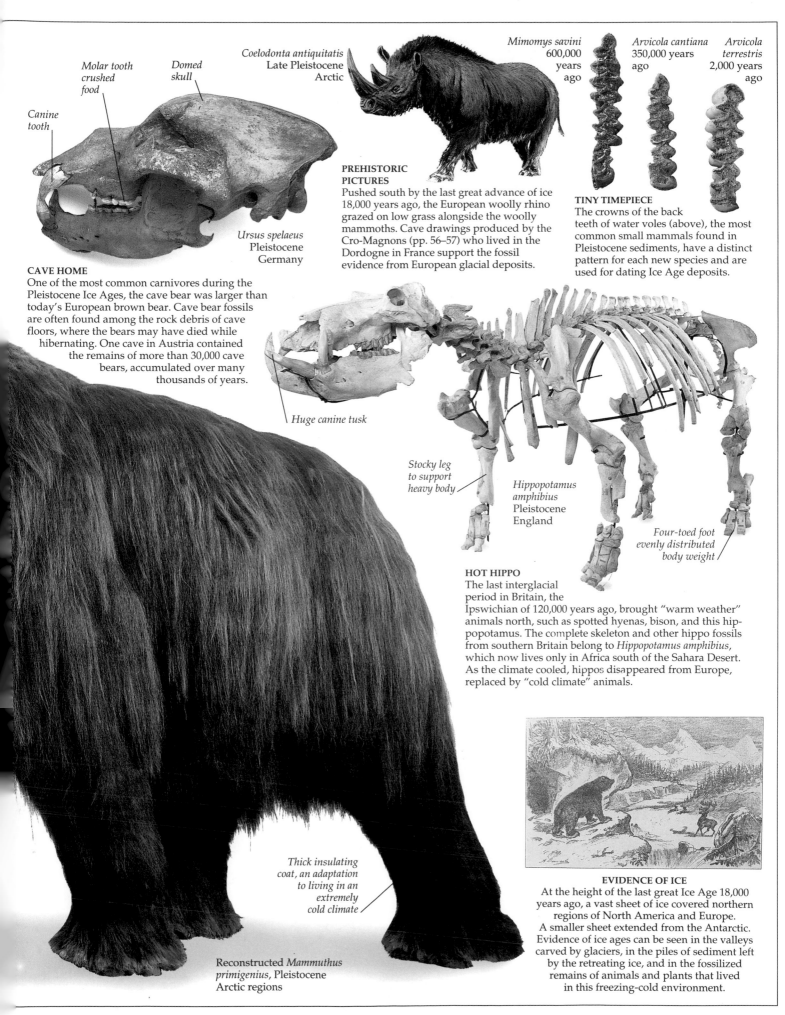

EVIDENCE OF ICE
At the height of the last great Ice Age 18,000 years ago, a vast sheet of ice covered northern regions of North America and Europe. A smaller sheet extended from the Antarctic. Evidence of ice ages can be seen in the valleys carved by glaciers, in the piles of sediment left by the retreating ice, and in the fossilized remains of animals and plants that lived in this freezing-cold environment.

Extinction

DEATH IS AS IMPORTANT as birth in the development of life. The extinction of one group of animals creates an opportunity for another group and, in a relatively short time, can completely change the direction of life. Mass extinctions have occurred several times in Earth's history. The largest extinction, at the end of the Permian period (248 million years ago), wiped out over half the marine families and two-thirds of the tetrapod animals on land. The main cause of the extinction of dinosaurs may have been a meteorite collision with Earth. The key lies in a layer of clay between rocks of the Cretaceous and Tertiary periods. Iridium, found in quantity only in meteorites, is so rich in this worldwide band of clay that, measured against the normal fallout from space, it could only have come from a massive meteorite.

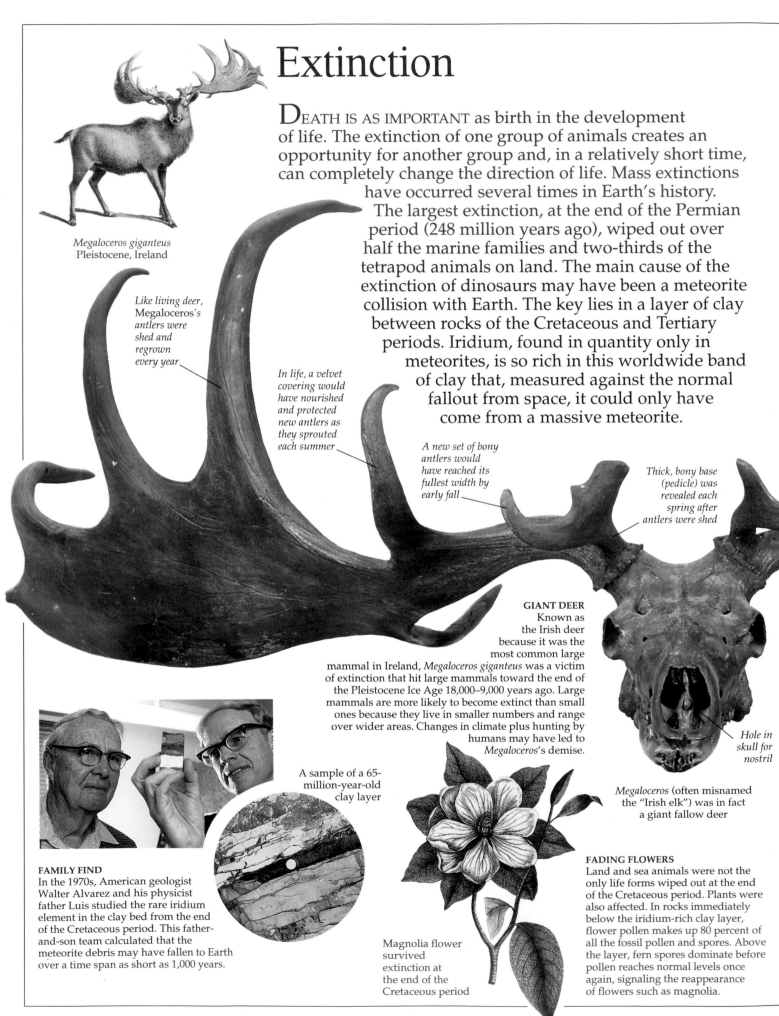

Megaloceros giganteus
Pleistocene, Ireland

Like living deer, Megaloceros's antlers were shed and regrown every year

In life, a velvet covering would have nourished and protected new antlers as they sprouted each summer

A new set of bony antlers would have reached its fullest width by early fall

Thick, bony base (pedicle) was revealed each spring after antlers were shed

GIANT DEER
Known as the Irish deer because it was the most common large mammal in Ireland, *Megaloceros giganteus* was a victim of extinction that hit large mammals toward the end of the Pleistocene Ice Age 18,000–9,000 years ago. Large mammals are more likely to become extinct than small ones because they live in smaller numbers and range over wider areas. Changes in climate plus hunting by humans may have led to *Megaloceros*'s demise.

Hole in skull for nostril

Megaloceros (often misnamed the "Irish elk") was in fact a giant fallow deer

A sample of a 65-million-year-old clay layer

FAMILY FIND
In the 1970s, American geologist Walter Alvarez and his physicist father Luis studied the rare iridium element in the clay bed from the end of the Cretaceous period. This father-and-son team calculated that the meteorite debris may have fallen to Earth over a time span as short as 1,000 years.

Magnolia flower survived extinction at the end of the Cretaceous period

FADING FLOWERS
Land and sea animals were not the only life forms wiped out at the end of the Cretaceous period. Plants were also affected. In rocks immediately below the iridium-rich clay layer, flower pollen makes up 80 percent of all the fossil pollen and spores. Above the layer, fern spores dominate before pollen reaches normal levels once again, signaling the reappearance of flowers such as magnolia.

Crater created by meteorite impact

Shock cloud from nuclear explosion

SHOCK CLOUD

The shock of a 6-mile (10-km) meteorite would have been greater than the explosion of all the world's nuclear weapons. Fires, tidal waves, and winds would have rushed around the globe. Dust thrown into the sky would have blocked the sun and interrupted plant life for many years. Meteorite collisions may seem rare, but satellite pictures show evidence of many craters. Research suggests that 90-mile (150-km) wide craters are made, on average, every 100 million years.

MAKING AN IMPACT

Meteorite impacts have left craters all over the Earth's surface, but many are obscured by plant growth and weather processes. This 210-million-year-old crater at Manicougan, Quebec, is about 42 miles (68 km) wide. The extinction at the end of the Cretaceous period may have been caused by a 6-mile (10-km) wide meteorite that left a crater 90 miles (150 km) wide.

Span of antlers was as wide as 11.5 ft (3.5 m), but body size was similar to today's North American moose

Antlers of rival stags locked in combat during rutting season every fall

DISAPPEARING DINOSAURS

At the end of the Cretaceous period (65 mya), dinosaurs rapidly died out, leaving their world to be taken over by mammals (pp. 44–49). In the slowly shifting continents and climates, dinosaurs were already in decline. However, over a relatively short time they disappeared forever, along with the large marine and flying reptiles, ammonites, and many brachiopods (pp. 16–17).

ROCK WRITING

Often found like pencil markings on rock, graptolites were once free-floating colonies of animals. The name comes from the Greek words "to write" and "stone." In life, graptolites were not flat lines but long branches of tiny cups. Inside each cup was a simple animal called a zooid, which lived in and was connected to the colony of other cup inhabitants. Graptolite colonies came in many shapes, including spirals, fans, and the twinned *Didymograptus*. Graptolites became extinct in the Carboniferous period, but some scientists believe that a similar animal still lives deep in the ocean.

Didymograptus murchisoni Early Ordovician U.K.

Flattened branch of cups

DEAD AS A DODO

Human interference can have a drastic effect on animals. Today, rare rhinoceroses, elephants, and tigers are threatened with extinction. The dodo, a large bird, lived on the then-obscure island of Mauritius in the Indian Ocean until the late 1700s. The arrival of pigs and other animals from passing ships disturbed the dodo's world, so that it became extinct about 100 years after its discovery.

Fossil finders

FOSSILS CAN BE FOUND in many places where rocks are exposed, such as quarries and coastlines, as well as in books and at museums. Collecting does not require much equipment; it is better to look among loose boulders than to hammer and chip at dangerous cliffs. At coastal edges, rising tides are a danger and tide time-tables must be consulted. Landowners must be asked for permission before fossil hunting begins. Some sites are so famous for their fossils that no collecting is allowed, thus preserving the treasures for everyone.

FAKE FOSSILS
Rocks often have interesting and curious shapes that can be mistaken for the fossil remains of organisms. One important reason why museums collect and study fossil remains is to compare them with new finds. Museums also help to identify fake fossils – stone or other material deliberately made to look like fossils. Some famous fakes have been produced, such as these sea creatures published in a book in Germany in 1726. Fake fossils are usually recognized eventually, but swindlers still try to sell them.

A collection of safety equipment and collecting tools for the young fossil finder

FOSSIL FOOTPRINTS
Some fossils, like footprints, cannot be removed from their enclosing rock. But a cast of the impression can be made in plaster and then removed. Footprints, as well as large fossil skeletons, can cover a wide area. A great deal of hard work may be needed to uncover a whole trackway, such as these dinosaur footprints.

Modern geological map

Sculptor's hammer

Geological hammer

Broad bladed bolster (special chisel)

Small rocks inset with brachiopods

FIELD TOOLS
Specially made geological hammers are useful for breaking open lumps of rock. Sculptor's hammers can be used with large chisels to split flat slabs of rock that contain fossils.

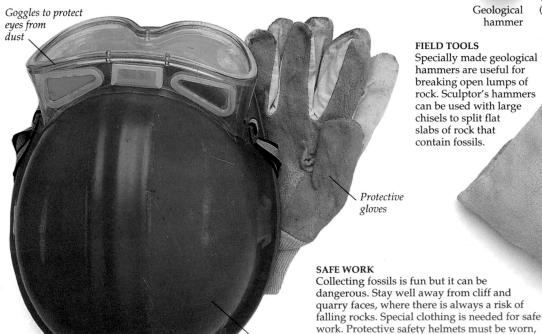

Goggles to protect eyes from dust

Protective gloves

Collecting bag for storing fossil finds

SAFE WORK
Collecting fossils is fun but it can be dangerous. Stay well away from cliff and quarry faces, where there is always a risk of falling rocks. Special clothing is needed for safe work. Protective safety helmets must be worn, and safety goggles will protect eyes from flying chips of rock when hammering. Tough gloves will protect the hands from sharp edges. Cloth collecting bags are useful for carrying fossil finds.

Safety helmet to protect head from falling rocks

Old geological map of rock
distribution on both sides
of the English Channel

ROCK LAYOUT
Only certain kinds of rocks contain
fossils, and it helps to know where
these rocks are. Just as there are maps
showing towns, mountains, and rivers,
geologists have produced maps that
show where different rock layers can
be found on the Earth's surface.
These can be studied to
discover where fossils
might be found.

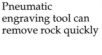

Pneumatic
engraving tool can
remove rock quickly

Diamond-edged
dental wheel
for cutting
into the rock

PREPARING FOSSILS
Sometimes fossils are
collected still embedded
in rock. Many tech-
niques and tools are
used to expose them.
Chemicals may be
used to dissolve the
rock. It may be possible
to brush or pick the rock
material from around
the fossil, but in hard
rocks, hammers and
chisels are used to chip
away the matrix, or
surrounding material.
Many months, or even
years, may be needed to
complete the work.

A small hammer
and chisel delicately
remove rock

FINE TOOLS
Small chisels will help when
working close to the fossil.
Brushes are used to dust
away fragments of rock.
Tough-bristled toothbrushes
and water will remove dirt
from robust fossil surfaces.
Broken fragments of fossil can
be carefully picked up with
tweezers before sticking
them together with
adhesives. A hand lens
may be used to view the
fossil close up. Photos and
notes of the work done
may help in the future
with similar fossils.

Narrow
chisel

Dusting
brush

Toothbrush

Fine
tweezers

KEEPING RECORDS
Many books, both old and
new, illustrate fossils found in
particular localities or show
the detailed differences among
similar looking specimens.
These books can be consulted
to help identify the fossils
found. Some fossils may be
very rare, so many books may
need to be checked before the
fossils can be identified.
Labeled drawings of your
fossils will help you recognize
the shapes and details.

Fossil
enamel
surface

Root missing

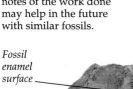

Molar
tooth of
Gomphotherium

Tooth
crown

FRAGILE FOSSILS
Fossils are often very
delicate and fragile. Some can be
stored quite easily in boxes, wrapped in
tissue paper, but others, like this molar tooth
of ancient elephant *Gomphotherium* (pp. 48–49), may
need some attention first. Liquid plastics may be painted on
the surface to strengthen the tooth by filling microscopic cracks.

Index

Acknowledgments

Dorling Kindersley would like to thank:
The Trustees of the National Museums of Scotland, Dr. Tim Smithson, and Dr. Michael Coates (University of Cambridge) for their valuable assistance on making the model of *Westlothiana*.
For help on research and photography: Dr. Ian Rolfe, Dr. Bobbie Paton, Bill Baird, Bob Reekie, and Dr. Sheila Brock, National Museums of Scotland; Dr. Neil Clark, Hunterian Museum; Dr. Ken Joysey, Dr. M. Coates, Dr. Jenny Clack, Ray Symonds, and Sarah Finney, Zoology Museum; Dr. Robin Cocks, Dr. Jerry Hooker, Sandra Chapman, Andy Currant, Valerie Harris, Peter Whybrow, Robert Kruszynski, Ann Lum, Helen Santler, and Tim Parmenter, Natural History Museum; Dr. David Norman, Dr. Simon Conway Morris, Dr. Barry Rickards, Mike Dorling, Rod Long, and

Margaret Johnston, Sedgwick Museum; Alastair Gunning and Patricia Bascombe, Glasgow Museum Services; P. Chadwick, A. Crawford, J. Downs, L. Gardiner, S. Gorton, C. Keates, D. King, M. Long, and A. Neimenn for extra photography.
For kind permission to photograph: *Acanthostega* and *Ichthyostega*, Geological Museum, University of Copenhagen, Denmark; *Sinokannemeyeria*, Paul Howard, Yorkshire Museum and Ian O'Riordan, Edinburgh City Art Centre; *Skakoper cryptozoan*, Leicester University Geological Department; Model of woolly mammoth, Royal British Columbia Museum, Victoria, Canada.
For design and editorial assistance: Manisha Patel, Sharon Spencer, Helena Spiteri, and Scott Steedman.
Model maker John Holmes
Artwork Simone End, William Lindsay, and John Woodcock

Picture credits
t=top b=bottom c=center l-left r=right
Bayerische Staatssammlung fur Palaontologie and Historische Geologie, Munich 41tr, 41br. Bettmann Archive 13tl, 14cl. Biofotos/Stames Summerhays 52c. Dr. Neil Clark 18cl, 21bl. Cleveland Museum of Natural History, Ohio 21tl. Dr. Michael Coates 25tr. Bruce Coleman Ltd 12tl, 17cr; H. Reinhard 22tc; Jane Burton 28tl, 43cr; V. Serventy 50br; John Canoclosi 62c. By permission of W. W. Norton & Co. Ltd. and Marianne Collins 14tr, 15tl. Simon Conway Morris 6c, 12c, 12bc, 12cr, 14cr, 14bc, 15tr, 15c, 15cl, 15bc. © Crown Copyright 62tr. Mary Evans Picture Library 20tl, 53tc, 59br. F.L.P.A.: S. Johnson 10tr, 44tr. Geological Society 38tl. Geological Survey of Greenland, Copenhagen: A. Garde 11br. Glasgow Museums 26–27b. Robert Harding Picture Library 10c. Dr. Gillian King, South African Museum: C. Booth 30bl;

R. M. H. Smith 31cr. Paul McCready, AeroVironment Inc. 40cl. © Trustees of the National Museums of Scotland 1994 22tl, 22bc. The Natural History Museum, London 8br, 13bl, 22tr, 30bc, 31c, 40tr, 43tr, 44br, 45tl, 46c, 47cr, 49tr, 50tr, 51tr, 57c, 58tl, 59tc. Novosti Photo Library 58bc. Oxford Scientific Films 52bl. Planet Earth Pictures/Peter Scooner 19tc. Science Photo Library 18tl; Ludek Pasek 24tl, 27tl; Martin Dohrn 41cr; John Reader 55tr, 55br, 56tl; Roger Ressmeyer Starlight 60cl; Prof. Walter Alvarez 60bc; N.A.S.A. 61tl; U.S. Navy 61tc; David A. Hardy 61tr. Forschungsinstitut and Naturmuseum Senckenberg, Frankfurt 44–45c. Paul C. Sereno 32tr. Wojciech Skarzynski 33br. Sternberg Museum, Hays, Kansas 20br, 21cr. Texas Memorial Museum, The University of Texas at Austin, courtesy Prof. Wann Langston Jr. 40–41bc. Dr. P. Wellnhofer 40cl, 40bl, 42bl, 43c. Dr. R. T. Wells, Flinders University, photographer F. Coffa 50cl.

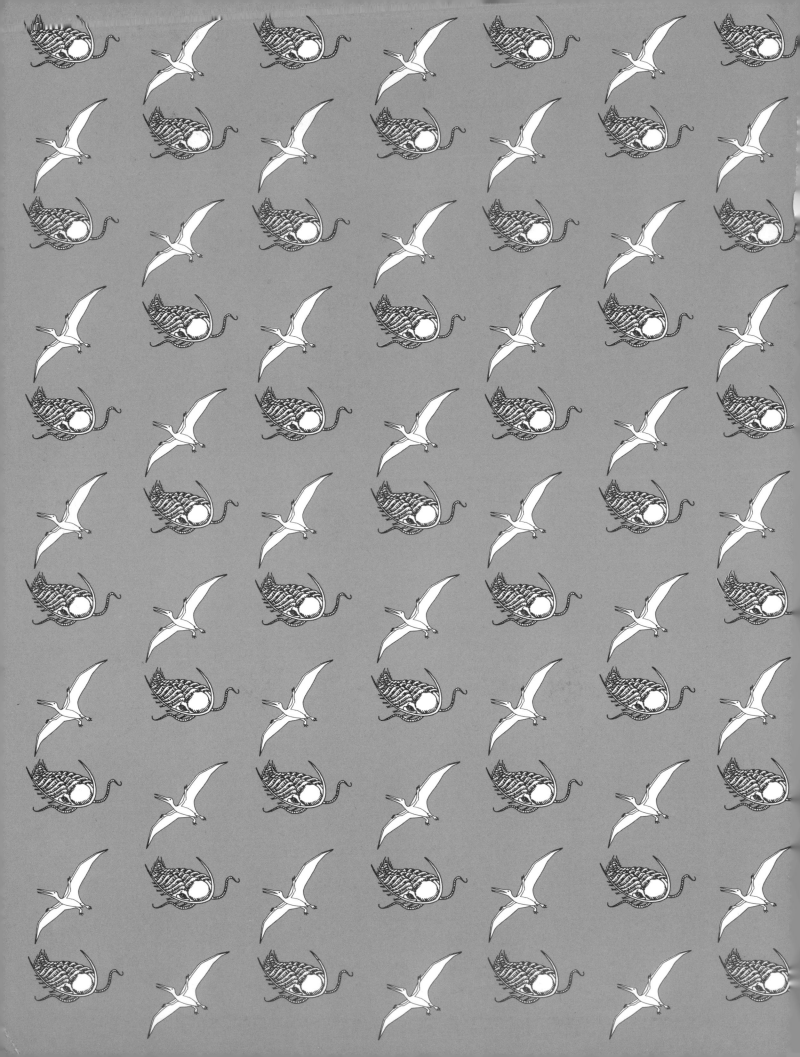